Recycling:
The Alternative
to Disposal

Recycling

THE ALTERNATIVE TO DISPOSAL

A Case Study of the
Potential for Increased
Recycling of Newspapers and
Corrugated Containers
in the Washington Metropolitan
Area

108662

THOMAS H. E. QUIMBY

*Published for Resources for the Future, Inc.
by The Johns Hopkins University Press
Baltimore and London*

RESOURCES FOR THE FUTURE, INC.
1755 Massachusetts Avenue, N.W., Washington, D.C. 20036

Resources for the Future is a nonprofit corporation for research and education in the development, conservation, and use of natural resources and the improvement of the quality of the environment. It was established in 1952 with the cooperation of the Ford Foundation. Part of the work of Resources for the Future is carried out by its resident staff; part is supported by grants to universities and other nonprofit organizations. Unless otherwise stated, interpretations and conclusions in RFF publications are those of the authors; the organization takes responsibility for the selection of significant subjects for study, the competence of the researchers, and their freedom of inquiry.

This study was prepared as part of RFF's quality of the environment program directed by Allen Kneese and Blair T. Bower. It was edited by Ruth Haas. The charts were drawn by Federal Graphics. Thomas H. E. Quimby, formerly associated with RFF, is now administrative assistant to Congressman Richard Vander Veen.

RFF editors: Mark Reinsberg, Joan Tron, Ruth B. Haas, Joann Hinkel

The Johns Hopkins University Press, Baltimore, Maryland 21218
The Johns Hopkins University Press Ltd., London

Library of Congress Catalog Card Number 74-6836
ISBN 0-8018-1655-6

Library of Congress Cataloging in Publication data will be found on the last printed page of this book.

Contents

Preface   ix

1. Introduction   3

2. Generation and Disposal of Residuals: An Overview   7

Introduction   7
The Interrelatedness of Residuals   8
Recycling as an Alternative to Discard   9
Classification of Residuals by Generating Source   10
Recyclability of Residuals   13
Content of Municipal Refuse   17
Measurement of Solid Residuals Collected   19

3. The Solid Residuals Handling and Disposal System in a Municipal
Area: Elements and Costs   23

Introduction   23
On-site Handling, Processing, and Storage   26
Collection and Transport   30
Collection of Residuals for Recycling   34
Processing and Transfer   36
Landfilling   39

4. Flow of Paper Through the Production–Consumption Sequence   43

Basic Varieties of Paper   43
Paper Products   44

Contents

Newsprint and Newspapers: The Products   47
Corrugated Containers: The Products   49
Fabrication of Corrugated Containers   50
Functions of Corrugated Container Board   51
End Uses of Corrugated Container Board   53
Residuals of Paper Production and Consumption   55
Converting Residuals   57
Distribution Residuals   58
Consumption Residuals   59
Commercial Consumption Residuals   60
Newsprint and Newspapers: The Residuals   61
Disposition of Newspaper Residuals   63
Generation and Disposition of Corrugated Container Board Residuals   64
Raw Material Qualities of Corrugated Container Residuals   66
Relative Cost of Paper Residuals and Virgin Pulpwood   69
Effect of Residuals Management on Relative Costs   73
Demand for Recycled Paper   75
Function of the Paper Stock Dealer   76
Concluding Comment   83

5. The Generation, Disposal, and Recycling of Newspaper and
Corrugated Container Residuals in Metropolitan Washington   85
    Introduction   85
    Newsprint Products and Newspaper Residuals in the DC/SMSA   86
    Factors Affecting Recycling   89
    Potential for Separate Handling of Newspaper Residuals   91
    Availability of Newspaper Residuals in Apartment Houses   93
    Potential for Increased Recycling of Newspaper Residuals in the DC/SMSA   96
    Generation of Corrugated Container Residuals in the DC/SMSA   97
    Summary   101

6. Effects of Recycling Newspapers and Corrugated Containers
on Solid Residuals Management Costs in the DC/SMSA   103
    Introduction   103
    Integration of Residuals Flows to Recycle and to Final Disposal   107
    Solid Residuals Management System I   108
    Solid Residuals Management System II   109
    Solid Residuals Management System III   118
    Comparison of the Three Systems   122
    Feasibility of Additional Recycling in the DC/SMSA   125
    Appendix   126

7. Concluding Comments   129

Tables

1 Some Reported Compositions of Mixed Solid Residuals Collected by Municipalities 18
2 Major Categories of Paper Consumption, 1969 46
3 Classification of Shipments of Corrugated Container Board by End Use, 1969 54
4 Approximate Quantities of Paper Residuals Generated, Recycled, and Discarded by Stage of Generation, 1969 61
5 Estimated Disposition of Newsprint and Newspaper Residuals in 1969 62
6 End Uses of Recycled Newsprint and Newspapers in 1969 65
7 End use of Recycled Corrugated Container Board Residuals in 1969 67
8 Paper Stock Grades Describing Newsprint and Newspaper Residuals 79
9 Quantity and Price of Grades of Corrugated Paper Stock Recycled in 1969 80
10 Materials Balance for Newsprint in Washington for 1969 89
11 Solid Residuals Quantities and Costs, 1969–70 for DC/SMSA 104
12 Total Annual Costs for Solid Residuals Management System I and Variations 113
13 Total Annual Costs for Solid Residuals Management System II and Variations 115
14 Total Annual Costs for Solid Residuals Management System III and Variations 120

Figures

1 Stages of residuals generation. 11
2 Major elements of a solid residuals handling and disposal system. 25
3 Stages of paper residuals generation and recovery of disposal. 56
4 Flow of newsprint and newspaper through the DC/SMSA, 1969. 87
5 Flow of corrugated container residuals through the DC/SMSA, 1969. 99
6 Flows of residuals to reuse or landfill in the DC/SMSA, 1969. 110
7 Solid residuals management system I. 111
8 Solid residuals management system II. 112
9 Solid residuals management system III. 119

Preface

The study reported here was undertaken at a time when the term "recycling" was just becoming popular. One position espoused was that recycling *per se* was good; if some recycling was good, more was better; "complete" recycling was best. However, at that time essentially no empirical studies had been carried out on the costs and consequences of different levels of recycling for any one type of residual, not to mention for the economy in general. Nor had the factors influencing the degree of recycling at any point in time been identified in detail. An increase in recycling was rationalized on the basis first, that it would reduce solid residuals management costs, and second, that it would decrease the drain on renewable and nonrenewable natural resources and the inputs associated with extracting and processing those resources.

This pioneer study was designed to address the first of these two rationales on an empirical basis. The approach was to take a major metropolitan area and a major component of the solid residuals generated in such an area—paper residuals—and analyze what elements of this class of residuals might be recycled, at what costs, the constraints inhibiting recycling, and the impact increased recycling would have on solid residuals management costs.

Quimby has set forth a clear exposition of the "systems" nature of the problem. He starts with the points in the production and use of goods where solid residuals are generated and describes the factors that affect the economics of handling and disposing of the residuals generated at those points.

This study makes it clear that recycling is an economic phenomenon. That is, the extent a given material is recycled is a function of the values of the so-called secondary materials in relation to so-called virgin materials. These relative values can change as a result of many factors, including changing technology, tax policies, or costs of other inputs. In fact, the value of secondary materials in the paper industry shows a substantial fluctuation. For example, in the three-year period during which the study was underway, the market price for used newspapers about doubled, reflecting to some degree the substantial increase in the cost of paper produced from virgin wood. However, by early 1975 the price had decreased again, at least for used newspapers, to about the level that existed at the time of the study. Of course the absolute costs of solid residuals handling and disposal, and of paper residuals processing have increased substantially because of inflation. But it is the relative price of the two alternative raw materials that determines the extent of recycling. This is the most important "message" of the study, along with the identification of the different factors that influence those relative prices.

Because of the doubling, tripling, and in some cases even larger increases in energy and fuel costs in recent years, more attention is being given to mixed solid residuals as a fuel substitute in thermal power plants or as a fuel in plants specifically designed to "recover energy" from the mixed solid residuals. Because the trend in energy costs will certainly continue upward, this alternative will become increasingly attractive. It should be emphasized, however, that, as shown in this study, cellulose fiber in the paper residuals of municipal solid residuals is becoming more valuable as a raw material for the production of paper products than as a fuel.

Quimby also points out that using secondary materials, such as paper residuals, for the manufacture of new products may not necessarily be a gain environmentally. The processing of residuals for raw material input results in the generation of residuals, just as does the processing of virgin materials. Here again, from society's point of view, what is needed is a "total system" approach—one that considers all activities, from forest management through reuse or disposal of the final product. Although the

ix

context of this study is limited to a single metropolitan area, it should be viewed in the perspective of a total system.

The findings with respect to recycling of paper residuals in a metropolitan area merit emphasis. Though paper in general comprises a substantial portion of municipal solid residuals generated, the readily available used newspapers and used corrugated containers not already recycled are a relatively small portion of the total solid residuals handled in a metropolitan area. Nevertheless, even with 1971 prices, a small but perhaps significant reduction in the cost of total solid residuals management for the metropolitan area could be achieved by the increased level of recycling considered "feasible" by Quimby. Given the increasing costs of disposing of solid residuals by incineration and landfill, and given substantially increased value for paper residuals, economic conditions, independent of government policy, are working to increase recycling. This is the case even though all of the external costs, such as air pollution, associated with current typical solid residuals disposal practices are not included in the costs of recycling. As the constraints on the assimilative capacity of the environment increase in severity, the swing toward recycling will be even more pronounced.

Quimby has performed an important service in sifting through solid residuals in order to obtain sufficient empirical data for this study on one type of solid residual in a metropolitan area. Further work of this nature would be helpful in providing a firm basis for policy decisions.

April 1975   *Blair T. Bower*

Recycling:
The Alternative
to Disposal

1

Introduction

During his lifetime the average American will use and discard total solid materials exceeding 600 times his adult weight. The collection and disposal of these materials in metropolitan areas has become a matter of increasing concern to local governments as well as federal agencies.

The many ways of disposing of these used materials in homes, factories, and offices are sufficiently similar and interrelated to be combined in one conceptual framework, here designated the solid residuals handling and disposal system. This system consists of the operations that go on at the site of use, the aggregation of the used materials for disposal, and the disposal operations. Materials discarded after use may follow one of two paths: they may be buried in the ground with or without burning or other processing, or they may be recycled into a new product. It is generally thought that the extent to which materials are recycled relieves the environment of the burden of disposing of those materials. This book is essentially a study of the two paths and the theoretical and practical reasons for directing used materials to one path rather than the other. The used material chosen here is paper; specifically, newspapers and corrugated containers. The metropolitan area from which data have been collected is the standard metropolitan statistical area (SMSA) of the District of Columbia.

3

For a variety of reasons, a standard metropolitan statistical area is the most suitable economic and social entity in which to study the recycling of used materials. With increases in population and the growth of transportation, cities no longer mark the boundaries of human settlement as they once did. Concentrations of population spill over city, county, and state boundaries. The standard metropolitan statistical areas, defined by the federal Office of Management and Budget to provide a geographical framework for recording economic and social statistics, are each made up of a county or group of contiguous counties that contains at least one city of 50,000 inhabitants or more. There are 243 SMSAs within the fifty states, based on the 1970 census. The SMSA is not a perfect demarcation of concentrated or urbanized population because of its dependence on county boundaries for its perimeters. A good deal of rural space is included in some SMSAs. However, it is a geographical unit on which a large amount of economic and social information has been collected, and it is the best available medium for studying and comparing factors of human consumption.

The standard metropolitan statistical area of the District of Columbia includes the District itself, Arlington County, Fairfax County, the cities of Alexandria, Fairfax, and Falls Church in Virginia, and Montgomery County and Prince George's County in Maryland. Between the censuses of 1960 and 1970, Prince William County and Loudon County in Virginia were added to the metropolitan area. The population of the DC/SMSA in 1970 was 2,861,123, making it the seventh most populous metropolitan area in the United States. The population density of the SMSA was about 1,200 persons per square mile, with a density of about 12,300 people per square mile in the city of Washington.

Which path used materials follow in an SMSA, recycling or landfill, depends in large part on the amounts of materials available in any one place and the costs of collecting enough of them to serve as raw material for a manufacturing process. An analysis of the flow of material through the production–consumption system indicates the stages of the system at which used and scrap materials are available, the likely magnitude of supply at each stage, and the conditions of the materials at the various stages. This analysis is made in chapter 2.

Chapter 2 also compares recyclable materials with other used materials and discusses some of the problems of measurement that arise in determining the solid waste output of a city. While there is a conscious effort in this study to avoid substituting jargon for information, there are several

4

words commonly used in discussions of solid waste management and recycling which are ambiguous or misleading. Chapter 2 includes some definitions to indicate the meanings of these words in this study.

Over the years, a system for handling the material discarded in city living has become quite standardized. Some of the elements of the system, such as collection and final disposal, are so obvious as to be self-evident. Other elements, such as on-site handling, processing, and storage, are not so readily identified. The way in which operations are performed in one element has great influence on the efficiency of operations in subsequent elements. In some instances the management of operations in one element is completely separated from the management of operations in another. In other cases a single management agency controls all the elements. The relationships between the elements are basic to the economic decision of how much recycling there can be. These elements, their relative costs, and their relationships are examined in chapter 3.

Paper is a surprisingly general term in our civilization. For a discussion of recycling of paper to be informative, it must identify the product specifically and indicate the relationship of the product to the paper family. Chapter 4 is devoted to the identification of newspapers and corrugated containers and to the investigation of the paths that paper follows through the production–consumption sequence in the U.S. economy. This chapter also includes a review of the current recycling status of paper on a national scale and examines some of the explanations for the statistics, including the institutions and arrangements that currently exist.

Chapter 5 focuses on the flow of used newspapers and corrugated containers through the standard metropolitan statistical area of the District of Columbia (henceforth designated in the text as the DC/SMSA). A materials balance is constructed for the 1969–70 flow of newspapers and corrugated containers through the DC/SMSA. On the basis of several residential and commercial factors, judgments are made of additional amounts of newspapers and corrugated containers that might efficiently be recycled.

The solid waste management system elaborated in chapter 3 and the paper recycling operations in the DC/SMSA discussed in chapter 5 are integrated in chapter 6 to determine how the system costs of the District of Columbia metropolitan area may be affected by various levels of additional recycling of paper and various modifications of collection and disposal operations. Three basic systems are explored—the existing system, a system that provides for recycling additional newspapers, and a system that provides for recycling both additional newspapers and additional corrugated containers. Within each of the basic systems, variations in unit costs and disposal operations are examined. Some conclusions are drawn about the responsiveness of the systems to various cost factors.

Chapter 7 concludes the report with a brief discussion of the factors that determine the economic feasibility of recycling. Techniques of resource recovery other than recycling back to original product are mentioned. There is an indication of some of the difficulties of translating a monolithic exhibit of system costs into the real world where costs and benefits are distributed among a number of agents in the solid waste management system.

2

Generation and Disposal of
Residuals: An Overview

Introduction

There is no waste in the natural world. The products or effects of any process in nature are inputs to other natural processes. The fallen leaves of deciduous trees contain the nutrients for future growth. The shed shell of the lobster provides a local supply of calcium for the skeletal needs of other animals. The activities of man—extracting, harvesting, manufacturing, eating—result, in part, in outputs not considered useful by man. These he has termed wastes. The outputs considered useful have a limit to their usefulness, and when that limit is reached, that is, when they are consumed, they too are termed wastes. An output considered waste under some circumstances may be considered a by-product (useful) in other circumstances. Or, output considered waste by one person may be considered useful by another person. To term material or energy *waste* is a parochial judgment. This ambiguity is avoided if we use the word residual to denote all outputs of human production–consumption activities. The outputs of production include the product and the residuals. In the consumption activity the product itself becomes a residual. It might be noted here that this is simply another way of expressing the laws of the conservation of energy and mass. The total quantities of energy and mass

7

(with the exception of nuclear reactions) remain constant; they cannot be created or destroyed. The incessant flow of solid residuals from production–consumption activities is frequently referred to as the solid waste stream. This phrase is used in this discussion to differentiate those solid residuals that do not get recycled.

The valid meaning of the word *consume* should be noted. The dictionary definition "to do away with completely: destroy," is not acceptable as a predicate for matter or energy. It is quite misleading to think that any fraction of matter has somehow been subtracted from the sum total by the act of consumption.

The relevant concept of *consumption* here is fulfillment of a human want, or provision of utility. The act of consuming does not lessen the *quantity* of matter or energy involved; it *may* alter the form of matter and energy, but not necessarily. The words *consume* and *use* are employed interchangeably in this discussion. A newspaper has been consumed when one has finished reading it; a pair of shoes has been consumed when one considers them "worn out," that is, they no longer provide the desired services. A corrugated shipping container has been consumed when the goods contained in it have been unpacked. These residuals do not disappear; something must be done with them.

The Interrelatedness of Residuals

Residuals occur in solid, liquid, or gaseous form, and any one waste product may at different times occur in all three forms. For instance, when a householder grinds food residuals in the garbage disposal unit and flushes them into the sewer pipe, solid residuals are transformed into liquid residuals; they may then be processed to settle out in a clarifier where they again become solid residuals. It is the practice in some localities to incinerate sewage sludge; at this point they become partially gaseous residuals.[1]

Handling and disposal of solid residuals cannot successfully be carried out in isolation from the systems controlling gaseous and liquid residuals.

---

[1] The class description of residuals refers to the system that bears or transports them and does not necessarily denote their physical state. The particulate residuals of incineration are considered solid if they are removed from the combustion chambers mechanically; they are considered gaseous if they are airborne out the stack. The dry remainder of ground-up food residuals is considered a liquid residual so long as it is water-borne; it would be a solid residual if placed in the kitchen compost pile.

8

This is illustrated by the anticipation in Washington, D.C., that when private incinerators are closed down to reduce air pollution, the amount of solid residuals to be handled and disposed of will increase approximately 15 percent, from 3,000 to 3,500 tons a day. Reducing the amount of residuals discharged to the environment in one form increases the discharge of other forms, unless the residuals are recycled.

Recycling as an Alternative to Discard

All residuals of the production–consumption process have only two possible ends. One is discharge to the environment; the other is reuse (reclamation or recycling). It is apparent from the process of residuals generation that the only ways to reduce the quantity of residuals produced by human activities are by reducing the amount of energy and mass input into these activities or by recycling the residuals generated by human activity.

Recycling has become a very fashionable word, with connotations of virtue. What does it mean? Many newspaper stories give the impression that recycling is the collecting of old newspapers, empty soda bottles, and aluminum beer cans to be used again in the manufacture of new products. This is a misleading impression. Recycling has to be more than collecting; the collection of materials is only one of the operations necessary to accomplish reuse. However, an accurate description of recycling is not easily arrived at. When it occurs in a dictionary, *recycle* is defined as "to pass again through a cycle of changes or treatment."

After deliberating for 8 months to produce a definition of recycled paper for inclusion in purchasing specifications, the General Services Administration of the U.S. government issued a definition on August 2, 1971 that avoided the use of the word *recycle*, thus indicating that there is some difficulty in establishing its meaning unambiguously. These problems of meaning will become more clear as we discuss the stages of production–consumption activities at which residuals are generated, whether the reuse of these residuals constitutes recycling, and whether the products or services for which residuals are used makes a difference in the meaning.

Consensus on a definition of recycling is not important here; what is important is to be aware of its possible meanings, and to be able to determine from the context in which the word is used what is actually meant. The following are possible denotations:

9

1. The reuse of products in the same capacity for which they were originally manufactured, e.g., returnable containers such as bottles, crates, pallets; antiques; used houses; secondhand clothes.
2. The processing of residuals to produce the same raw material used in the initial manufacture of the final product. This meaning applies equally to paper, glass, or metals.
3. The alteration of the basic material of the residual to a completely different kind of material, e.g., using bacterial action to change cellulosic fibers in paper residuals to protein.
4. The release and use of the energy contained in the residual, e.g., generating steam for the production of electricity by incinerating solid residuals.

Here, the second of the four meanings is used, that is, recycling is the processing of used paper to make new paper products.

While recycling is the only alternative to discharging material to the environment, it may not always be the more desirable course of action in terms of environmental quality. Some recycling operations may result in the generation and discharge of residuals that are more damaging to the environment than the residuals of production—consumption activities the first time around. For example, the suspended solids generated in recycling paper residuals to produce a given product may exceed by 100 percent or more the quantity generated in producing the same product from virgin material. The amount of energy required for recycling may be greater than that required when using virgin material for a particular output. This is only to say that the benefits of recycling are not universal and incontrovertible; they must be examined on a case-by-case basis.

Classification of Residuals by Generating Source

For purposes of analyzing handling and disposal, residuals may be classified according to the stage of production—consumption activity at which they are generated. Figure 1 presents these stages in the manufacture of paper products. Only product material inputs and residuals are shown in this diagram. In reality there are processing energy and material inputs as well for every activity box, and there are residuals of process energy and materials issuing from every box. These are not shown in the diagram.

10

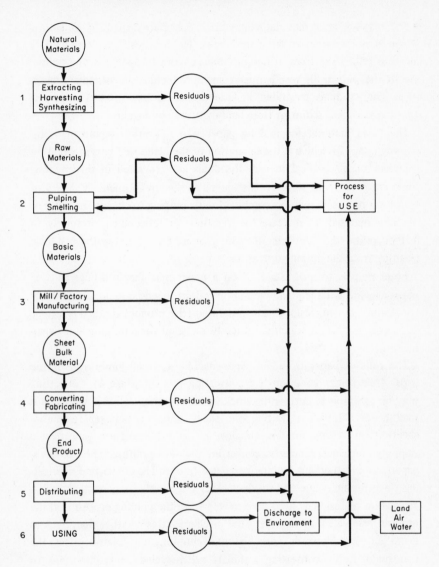

FIGURE 1
Stages of residuals generation.

The major natural material from which paper pulp is made in the United States is wood harvested from the forest. The harvesting process (stage 1) includes felling the trees, trimming limbs, cutting into lengths, and removing to the pulp mill. Woods operations may include debarking and chipping. The residuals generated include branches, leaves, bark, chips and sawdust as well as sediment from land disturbed by logging.

The basic material for making paper is the furnish or pulp made by pulping (stage 2), which is the separation of the cellulose fibers in wood by chemical or mechanical means. Residuals are generated in this process; some are reused and some are discharged to the environment. A variety of bleaching processes may be combined with pulping to change the color of the raw material, to increase its brightness or reflectance, or simply to further refine the cellulose. Bleaching operations increase the residuals resulting from the pulping operation.

Wood pulp is formed (stage 3) on a paper machine into huge rolls of paper as wide as 10 feet and wound on rolls at speeds as great as 5,000 feet per minute. Again, residuals are generated (the product residual is termed mill broke), most of which are directly recycled for repulping and paper-making.

The rolls of paper from the paper machine (termed jumbo rolls in the paper trade) next go through a converting process (stage 4). Converting may be as simple as cutting the stock from the jumbo roll into sheets, or it may be as complex as fluting one sheet, gluing it between two other sheets, then cutting, binding, slotting, and gluing it to form a corrugated shipping container. Similarly, converting includes printing paper for newspapers, or printing and binding paper for books. The converting–fabricating stage produces the end product. This stage is the first point at which generation of residuals is so far removed from the pulping process that the product residuals may flow into the discharge stream rather than back to manufacturing. However, with present technology and relative prices, the vast majority of converting residuals are recycled. It is important to emphasize that conversion–fabrication may or may not take place at the same location as basic materials manufacturing.

The distribution stage (5) of production–consumption activity also generates residuals. Distribution residuals may be products damaged in transportation, or more generally products left in inventory which are no longer salable. Durable items of some cost are not apt to become distribution residuals. Inexpensive products of only topical value, i.e., newspapers, always become distribution residuals.[2]

The use stage (6) of residuals generation can be divided into residential consumption residuals and commercial consumption residuals. Products consumed in homes become residential residuals; products consumed by stores, institutions, or factories result in commercial residuals. In all of these activities there exist the two possibilities of recycling residuals or discharging them to the environment. The course taken is a function of the generating source, economics, technology, and environmental management policies.

Recyclability of Residuals

The extent to which residuals are recycled depends on the cost of residuals as a raw material input into production and the cost of processing them compared with the cost of virgin materials as a raw material input and their processing cost. The factors affecting the prices of virgin raw materials are not considered here. We are concerned solely with the characteristics of residuals which affect their relative value as raw material inputs.

Characteristics of any substance considered as a raw material input— whether it be iron ore, bauxite, pulpwood, or waste paper—are mass, level of contamination, homogeneity, and location.

The significance of mass as a factor in recycling is readily seen when recycling is perceived as a sequence of operations involving materials handling, cleaning, processing, and forming. A certain level of aggregation or accumulation of volume is necessary before a material can be economically handled. Two thousand 80-page copies of the *New York Times* each weighing 1 pound and each in a different household have no value. But 2,000 of these tied together in a transportable, storable bale have a value of $18 or $20. Mass is a critical factor in determining whether materials of marginal value should be recycled or discarded.

Level of contamination is a second important characteristic in using residuals as raw material. Contamination refers to the extent to which the desired material occurs in combination with other materials and the difficulties of separating them. Contamination may arise in the production

[2] For a more detailed explanation of distribution residuals, see page 58.

13

process, during handling as a residual, or both. Cans with tin-plated steel sides and aluminum ends are an example of a material so adulterated that it cannot be recycled economically. Contamination occurs at the residential consumption stage when residuals not impervious to other materials are mixed for handling. The possibility of recycling paper residuals after they have been mixed with other household or commercial residuals is small because of the difficulty of separating paper from the grease and dirt in the flux of mixed solid residuals. If the contamination or adulteration of materials is specific, known, and limited to one type, the possibility of recycling is increased because specific operations for processing the residuals can be prescribed.

Homogeneity is another factor bearing on the recyclability of residuals. It is important as a raw material characteristic both for ease of processing and for quality of output. An example is the color of glass residuals. Clear glass and colored glass must be kept separate to provide a reasonably economic raw material for recycling. Residuals from the manufacture of folding boxes will downgrade residuals generated in the manufacture of envelopes if the two are mixed together. The greater the homogeneity of the residuals, the greater the potential for recycling.

The fourth characteristic of a used material which may increase or diminish its potential for recycling is its location. Location is significant in two different senses. In one sense it may refer to the residuals generating sites, which may be widely dispersed, such as detached dwelling units. A recyclable mass is developed by transporting the dispersed residuals to a central point. Location in this sense refers to the dispersion of residuals and implies a degree of effort to collect them in a recyclable mass. This aspect of recyclability is dealt with in the section on mass. The second sense in which location is a significant characteristic of residuals is the distance of the generating area or central collection point from a manufacturing plant equipped to recycle residuals into new products. It is in this sense that location is considered the fourth characteristic in determining recyclability.

The suitability of location is expressed in the cost of transporting residuals from generating area to manufacturing plant and transporting the finished product to the marketplace. Paper mills equipped to recycle used paper products have generally been located adjacent to urban areas, which are the sources of the raw material. This has provided an advantage both in the shipment of raw material to the manufacturer and the shipment of finished goods to the market. However, only about 20 percent of U.S.

paper mill capacity has been equipped to process paper residuals; the other 80 percent is located next to the forests, away from the urban areas. Because of the absence of a nearby manufacturing capacity, many of the paper residuals generated in urban areas do not have a good location characteristic.

The physical characteristics enumerated—mass, level of contamination, homogeneity, and location—are related to the different generation stages of the production—consumption activities shown in figure 1. The residuals of the pulping, smelting stage (2) are generally so similar to material input and output that there is no question of recycling them. In effect, they can be made compatible with raw material input and are on location for recycling. It is generally less costly to toss the residual back into the pot than to discharge it to the environment.

Whether to recycle or to discard residuals first becomes a relevant question when converting is carried out at a location different from manufacturing. Under current economic conditions and technology, recycling is generally more efficient than discarding. Converting residuals have the desirable characteristics of mass, homogeneity, and known contaminants. The processor who takes the residuals from a specific fabricator knows exactly what materials he will have to deal with. For example, the scrap generated in cutting polyethylene-coated board used for milk cartons commands a price of over $50 a ton from a paper mill which has a process especially adapted to stripping the plastic from the paperboard.[3] While treatment for contamination calls for special technology, the residuals are amenable to the treatment because of the specificity of the contamination and the homogeneity of the paperboard fiber. The large-scale operations of most conversion—fabrication processes assure a large quantity of residuals, fulfilling the recycling requirement of desirable mass. In conversion residuals, the characteristics of mass, level of contamination, and homogeneity are, by the nature of the process, favorable ones. It is conceivable that location would be so unfavorable that transportation costs would rule out recycling.

The extent to which the desirable aspects of mass, homogeneity, uniformity of contamination, and location occur at the distribution stage (5) varies. Homogeneity and uniformity of contamination are generally present because the product material is in the same form as it left the

---

[3] Author's conversation with representative of Riverside Paper Co. at New York City Chamber of Commerce recycling seminar, October 20, 1971.

converting plant. Mass may be achieved by the returns policy of the distributor who collects or credits the return of unsold goods, thereby providing a means of aggregating dispersed residuals into a mass. Location may have a positive quality if the distributor is well situated for loading and transporting residuals to the place of manufacture.

The physical characteristics of consumption residuals are such that a much smaller percentage of them are recycled in the current market than converting and distribution residuals. Consumption residuals are generated in both commercial (including industrial) and household activities. Economies of scale are available for the residuals generated in some commercial operations and in multiple dwellings. For example, the aggregation of newspapers from 300 apartments in one building presents a picture of costs quite different from that for the aggregation of newspapers from 300 single-family houses. The availability of 6 tons of IBM cards a month from a bank building is economically attractive. In general, the quality of mass is higher for commercial consumption residuals than for household consumption residuals. Homogeneity and uniformity of contamination are highest for converting residuals, intermediate for commercial residuals, and lowest for residential consumption residuals.

The characteristics enumerated—mass, uniformity of contamination, homogeneity, and location—are all physical characteristics. As is obvious in their delineation, they derive their value from the economics of the situation; that is, they are important to the extent that they contribute to the value of the residuals as a raw material compared with virgin material.

Competition between virgin material and residuals as raw material is affected by these qualities and also by: (1) the technologies of the raw materials processing and the associated costs for residuals and virgin materials; (2) the residuals management (pollution control) costs entailed in each technology; (3) the relative costs of transportation (both of raw material and processed material); and (4) the relative costs of the technologies of the two manufacturing processes.[4]

The relative demand for residuals compared with virgin materials may also be profoundly influenced by the structure of the manufacturing industry, particularly the integration of operations concerned with virgin material production. Demand may also be influenced by noneconomic factors, such as purchasing specifications that require virgin material but make no reference to performance characteristics of the final product.

[4] Blair T. Bower, *Economics of Residuals Use,* statement before the Subcommittee on Fiscal Policy of the Joint Economic Committee of the Congress, November 9, 1971.

16

Content of Municipal Refuse

The composition of mixed solid residuals (MSR) collected in munici-palities—what kinds of materials and how much of each—is a matter of interest, both for indications of the materials and proportions that might be amenable to recycling, and for clues on the kinds of processing that would help reduce the volume of MSR going to landfill. It is also helpful in designing materials handling systems to know how mixed solid residuals are measured and what the relationships are between weight and volume.

At first glance it should be easy to determine the annual national total of municipal mixed solid residuals by adding up the production figures of industry and making adjustments for goods added to inventory. The total minus additions to inventory should equal what is discarded as waste. However, this is not a valid method for several reasons: production statistics do not include nonproduct output, which comprises a large portion of residuals to be disposed of; mixed solid residuals collected in a city may contain as much as 30 percent grass clippings, leaves, and yard waste; finally, many residuals may be discharged to the environment at site of generation by burning or being flushed away. Thus total mixed munici-pal solid residuals cannot be accurately derived from product output.

Table 1 shows four analyses of the composition of municipally collected mixed solid residuals. They are presented to illustrate the variations in mixed solid residuals caused by a variety of factors. Geography, local disposal practices, income, local consumption patterns, the season of the year, the local industrial base, and activity mix (proportion of household residuals collected compared to commercial residuals collected) result in wide variations of municipal MSR composition. The techniques used for sampling and analysis also affect the final result.

There are many reasons for the different analyses shown in table 1. For example, private haulers rather than public agencies collect 40 to 60 percent of solid residuals generated in many cities. The private haulers collect from commercial establishments and multiple dwelling units, result-ing in MSR of a composition different from that collected by the public agency in the same area. Another reason may be the different treatment of moisture content in the sampling process. The analysis of residuals in column A of table 1 is described as being on a dry basis. The other three columns include in the quantities assigned to each residual the moisture that was present at the time the sample was taken. In the case of column B, the sample taken at the incinerator in Alexandria, Virginia, the moisture was assumed to be 20 percent. Obviously, it is necessary in assessing the

17

TABLE 1
Some Reported Compositions of Mixed Solid Residuals
Collected by Municipalities

Residual	Percentages by weight			
	A[a]	B[b]	C[c]	D[d]
Paper products				
Newspapers	10.3			
Corrugated containers	23.9			
Other Paper	21.8			
Total paper	56.0	58.0	17.5	40–54
Yard and garden waste	7.6	8.4	13.5	3–80
Wood and Christmas trees	2.5	1.4	0.9	3–70
Dirt and debris	3.4	3.4	0.3	1–50
Plastics, rubber, leather	4.3	3.3	2.3	1–20
Glass and ceramics	8.5	8.1	17.9	8–11
Food wastes	9.3	6.1	32.6	10–26
Rags and miscellaneous	0.9	3.1	0.5	1–20
Metals	7.5	8.2	14.5	8–11
Total	100.0	100.0	100.0	

[a]Adaptation of a tabulation of typical refuse analysis by E. R. Kaiser, "Chemical Analysis of Refuse Components," in *Proceedings 1966 National Incinerator Conference,* New York, May 1–4 (American Society of Mechanical Engineers), pp. 84–88.
[b]A sample of municipal refuse taken and measured at the Alexandria, Va., incinerator in May 1968. From A. C. Achinger and L. E. Daniels, "An Evaluation of Seven Incinerators," in *Proceedings of 1970 National Incinerator Conference* (U.S. Environmental Protection Agency), p. 36.
[c]Annual average composition of domestic refuse collected in Flint, Michigan, based on samples taken in June and January; from *Solid Waste Disposal Study, Genesee County, Michigan* (U.S. Dept. of Health, Education and Welfare, 1969), p. IV–7.
[d]Typical percentage ranges of the components of municipal solid wastes; from H. Lanier Hickman, "Characteristics of Municipal Solid Wastes," in *Scrap Age* (February 1969) pp. 305–307.

burden of any solid residual on the disposal system, or in comparing residuals generation in different areas, to know how the moisture content was assessed. Metals and glass are not affected by moisture, but their proportions increase when their percentages are calculated on the basis of dry weight and decrease when the percentages are calculated on the basis of atmospheric or load moisture. Absorbent fractions such as paper and textiles acquire weight from the moisture-rich fractions of food wastes.

The substantial difference between columns B and C in the paper fraction may be due in part to the difference in per capita income between Alexandria, Virginia, and Flint, Michigan, but it is probably due primarily to the backyard burning practiced in Flint but not in Alexandria. Another factor is that the refuse sampling in Flint covers only residentially generated residuals, while that for Alexandria includes both residential and commercial generation. It is important to keep in mind that all of these figures are based on residuals *collected,* including those from street cleaning, not residuals *generated.*

The quantities of yard and garden waste collected in municipal MSR vary seasonally as well as geographically. The range in Flint, Michigan, went from a high of 26.7 percent in June to a low of 0.3 percent in January to yield the 13.5 percent reported in column C. Column D shows this category, with a 3–80 percent range, as being the most wide-ranging of the MSR fractions. Metals and glass showing a range of 8–11 percent in column D may be reflecting the absence of moisture in the sampling operation as well as a remarkable average consumption. The percentage of food wastes in Flint MSR, 32.6 percent, compared with the food wastes in Alexandria MSR, 6.1 percent, may reflect the presence of garbage grinders in Alexandria and their absence in Flint.

Measurement of Solid Residuals Collected

The statistics of production—how many tons are produced, how many gross are shipped, how many cords of wood are cut or barrels of crude petroleum pumped—tend to be hard statistics recorded in tallies of production, on bills of lading for goods shipped, and on invoices for goods sold. In contrast, the data relating to solid residuals management do not have this kind of firm foundation. No one dealing with them should accept them with the same certainty that he accepts statistics of production and of raw material inputs to production.

There are two factors needed for the development of accurate statistics on MSR—a universal unit of measurement and a tool for measuring. These are present only in some of the situations in which MSR are being handled. Should the unit of measurement be cubic yards, or tons? Volume rather than weight is the limiting factor in a substantial portion of the collection truck activity. Volume is also the critical factor in landfilling operations, so that the cubic yard would be a more informative unit of measurement than a ton. The problem is to know when a cubic yard of mixed solid residuals will remain the same. It varies according to the compressibility of the residuals and the compression applied. The compressibility of the residuals varies with their composition. A loose cubic yard of mixed solid residuals—that is, the amount that would be held by a 1-yard portable bin without special compression—is considered by some MSR haulers to weigh 180 pounds on the average, but it may fall to 50 pounds or lower and might go to 200 pounds. Packer trucks achieve densities of 400 to 600 pounds per cubic yard in their loads. Mixed solid residuals compacted in a landfill generally achieve a density of 1,000 pounds per cubic yard. In the absence of weigh scales, statistics on the quantity of MSR handled are developed by visually relating the load on a truck to the volumetric capacity of the truck. This is not an accurate procedure.

Most records of solid residuals management are kept in tons; there is less ambiguity in these figures. But the information they yield about solid residuals handling must be supplemented by knowledge of densities of materials composing the residuals. For instance, the compacted density range of papers runs from 1,200 pounds to 1,900 pounds per cubic yard; the density of wood ranges from 300 to 1,900 pounds per cubic yard; glass ranges from 1,800 to 4,900 pounds per cubic yard; and copper alloys from 13,500 to 14,850 pounds per cubic yard. If the variation in density of different materials is kept in mind, solid residuals information expressed in tons becomes more useful.

The second factor often lacking in statistical data on municipal handling and disposal of solid residuals is the measuring device, the scale. In the past, with an apparent abundance of low-cost land available nearby for dumping, investment in truck scales and weighing personnel was not considered necessary. Information on the quantity of residuals being discarded was not necessary for the solution of any problem. When the city fathers decided that dumping fees should be assessed, fees were based, and tallies were kept, on the basis of cubic yards of truck capacity and whether or not the residuals were loose or compacted. When information was required for municipal budgets, total annual solid residuals disposal in a particular jurisdiction was computed on the basis of an estimated per capita per calendar day generation multiplied by the population. The per capita figure was derived either from a similar jurisdiction using a scale, or from some sort of sampling. This system of compiling data on solid residuals management activities obtains today in many localities.

Some notice has already been taken of the significance of moisture content for the statistics of municipal MSR management. Moisture does

20

not increase the volume of solid residuals; it only adds to their weight, depending on the sorbent qualities of the residuals. The contribution of moisture to the weight per unit of volume could be higher than 50 percent for paper and textile residuals; it would not contribute any weight to a unit of glass or metal.

3

The Solid Residuals
Handling and Disposal
System in a Municipal Area:
Elements and Costs

Introduction

The elements of a solid residuals handling and disposal system are the operations that take place between the point of residuals generation in the production–consumption sequence and the point at which the residuals are deposited in the natural environment. The system may be complex, encompassing advanced technology and requiring substantial inputs of energy, as in the steel and paper industries, or it may be as simple as a littering picnicker who deposits his empty beer can on the ground as an act of disposal at the point of consumption.[1] The management of the system may lie in the hands of a single agent, as in the case of a farm household that buries its residuals within the boundaries of the farm or as

---

[1] It should be noted that *final* in *final disposal* refers to last human handling. The can may be at its point of discard for 500 years before it is assimilated into natural cycles. *Disposal* simply denotes transfer from the human sphere of activity to the natural sphere of activity.

in the case of large industry that performs all operations to dispose of its solid residuals. Or the management may be divided, as it often is for urban activities, so that one agent is responsible for processes at the generating site, a second agent is responsible for collecting and transporting the residuals, and a third agent is responsible for depositing them in the environment. This discussion is concerned with residuals handling and disposal in a municipality where there are several agents involved in the management of residuals.

While municipal solid waste generally consists of the residuals that come from households and commercial activities, it may also include manufacturing residuals from smaller industries that do not find it economical to maintain and operate their own disposal facilities. The extent to which municipal solid waste contains industrial residuals is more a matter of the size of the industrial plant and the city than a difference of function.

The elements of a municipal solid residuals handling and disposal system are shown in figure 2. The sequence of operations from production–consumption to final deposition is: 1, on-site handling, processing, and storage; 2, collection and transport; 3, transfer and processing; and 4, landfill.

Residuals, once generated, may flow through the system to final disposition in the environment, or they may be diverted to recycling at any step along the way. Whether or not they are diverted is a matter of the costs of the operations involved and the value of the residuals.

The costs of the various elements of the solid residuals handling and disposal system may be priced, unpriced, internal, or external, and will be referred to by these terms. Priced costs are readily identified as the charges that are tallied in the bookkeeping of any enterprise. Unpriced costs are not part of a profit-and-loss statement, but nonetheless they represent a transfer of value. Examples of unpriced costs are the labor of a householder in putting his trash containers out for collection, and the reduction in the populations of terns preyed upon by seagulls who have proliferated as a result of feeding in open dumps on the eastern seaboard.

Internal costs are costs the presence and magnitude of which affect the system, e.g., the costs of paying labor or operating equipment. External costs are imposed by the system on an element outside the system with no penalty or reward resulting for the system. For example, the costs of operating a landfill may not be increased if paper blows from the dumped refuse. But the scattered paper debris will depress the value of adjacent land. This study is concerned primarily with priced, internal costs, but significant unpriced, external costs will be noted where they are relevant.

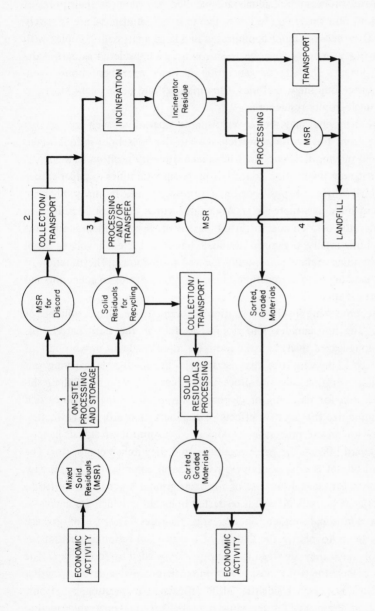

FIGURE 2

Major elements of a solid residuals handling and disposal system.

On-site Handling, Processing, and Storage

In on-site processing, residuals are handled or stored at their point of generation. In a single-family house the activities of disposal are relatively simple; they become more complicated in a large apartment complex with perhaps thirty or forty separate trash rooms, all considered as part of the on-site system. The site of generation includes apartment house complexes, office buildings, factories, supermarkets and department stores, as well as single-family residences.

On-site handling may be a very simple operation, such as carrying residuals from the bedroom wastebasket to the household residuals container and moving that container from its customary position to the street curb, or it may involve the collection by pneumatic tubes of paper wastes from thirty different positions in a printing plant. In many high-rise apartment buildings, there are gravity chutes which guide solid residuals to an incinerator or a compactor in the basement. Some garden apartments have a horizontally organized residuals handling system involving small tractors pulling trailers to centrally located trash rooms. The function of on-site handling is to aggregate residuals for storage prior to collection and transport off site.

On-site processing involves alteration of original size, shape, volume, or density. The householder who stomps his foot in the trash barrel (this takes a long-legged man) to make room for more residuals, or who flattens his tin cans, or who ties his newspapers in bundles is carrying out processing operations. On-site processing reduces the volume of the residuals collected for recycling or disposal. On-site processing is also carried out to minimize the negative effects of residuals on health and aesthetics. Facilities for on-site processing of domestic consumption residuals include the backyard burner, the garbage grinder, and the kitchen compactor. The backyard burner is disappearing even from rural communities in response to demands for cleaner air. Use of a garbage grinder tends to be limited to more affluent households and is restricted in localities which are unable to take an additional burden on sewerage facilities. The use of garbage grinders tends to reduce the density of mixed solid residuals put out for collection since they are free of the moisture provided by food scraps. The kitchen compactor is the most recent domestic processing device. Whether this is an advance in residential MSR processing is questionable. Home compaction diminishes the recycling potential for the recoverable fraction of mixed solid residuals.

Reduction of the volume of MSR by baling, compacting, or incinerating is an on-site process which occurs more frequently in commercial and

26

industrial establishments than it does in single-family residences. In most densely settled areas, private incineration is now outlawed because of air pollution problems. Baling and compacting are becoming more widespread. These processes reduce the amount of space required for on-site storage, reduce the amount of labor required for collection, and reduce the volumetric capacity of the truck required for transport.

On-site storage of residuals takes place in two, three, or four phases of the handling, processing, and storage operations, depending on the complexity of the activities of the site and the means by which residuals are collected. In a simple household operation there are generally three phases—the wastebasket in the room where the residual is generated, the large trash container by the kitchen door or in the garage, and the curb location from which collection is made. In a factory there may be a single storage phase, as in the case of a dust collector that pneumatically draws in the dust generated by a metal polishing wheel and deposits it in a hopper from which it is loaded into a collection truck by gravity.

The nature and phases of on-site residuals storage can be integrated with handling and processing activities. For example, one on-site system in high-rise apartment houses involves handling residuals by a gravity chute from the floor of generation to the basement, the reduction of volume by a compactor at the bottom of the chute, and storage in a portable bin which, when full, is moved on casters to a point where it may be tipped into a refuse collection truck.

Space and facilities for on-site handling, processing, and storage of solid residuals do not receive the attention of architects, builders, and engineers at the design stage in the same way as do the front entrance to a house, the aisle layout of a department store, or the circulation of an office building. This tends to be most apparent in low-income housing where the amount of space available for residuals storage is relatively small and the population density is relatively high, resulting literally in a spilling over of stored residuals. It is least true of a manufacturing plant where the efficiency of space utilization is readily reflected in the costs of operation. Inadequate storage space, by interfering with production, increases internal production costs.

A major cause of urban litter is the inadequate design of on-site handling, processing, and storage facilities. A not unusual routine is for a householder to put his MSR containers out on the curb for collection on the appointed day. If the routine is to have setup men preceding the collection truck to remove container lids, there is a period of time, on a windy day, when the mixed solid residuals in the top of the container will

27

become airborne for some distance. This may be characterized as a failure of the collection system. It indicates that on-site storage design has not been correlated with the collection technique.

Whether the costs of on-site handling, processing, and storage are calculated or not, whether they are internal or external, depends to a great extent on the kind of institution at the site—whether it is a household, a drugstore, an office building, or a factory—and how efficiently it is organized. It also depends on the extent to which environmental quality standards are imposed on on-site solid residuals operations. The costs of on-site handling, processing, and storage of residuals are generally not calculated for detached residences. Sometimes they are perceived in political rather than economic terms. An example is the debate which took place in many municipalities a decade or so ago on the merits of backyard collection versus street-side collection of residential MSR. The debates were resolved mostly by political decision rather than by comparing the cost to the householder of moving his MSR to the curb with the cost of pickup of the MSR by the collection crew at the back door. Every segment in the on-site handling, processing, and storage of MSR has a cost. Some segments are readily priced, such as the capital, operation, and maintenance costs of a garbage grinder. Other segments, such as moving an MSR container from the back door to the street, are not so readily quantified.

Some costs are clearly internalized in on-site handling, storage, and processing, and some remain external. The cost of the householder's residuals container is borne by him, but the cost of paper blowing from his container into the yard of a neighbor on a windy day is an external cost that could be internalized by requiring that containers be filled to a maximum level.

The potential externalities of on-site handling, processing, and storage of residuals result in a very different set of sanctions than usually prevails in the provision of services to the public. The customary sanction for nonpayment of service charges is to shut off service. If a householder does not pay his water, light, or gas bill, the supply of water, light, or gas is discontinued. The damages or costs of discontinuance are borne directly by, and only by, the householder. The same sanction has been proposed for the provision of MSR collection service to householders. Consideration of the potential externalities involved should make it apparent that this is not a workable sanction. The householder immediately affected will bear the damage of discontinuance of service, but so will all the adjoining householders.

28

The internalization of costs by means of environmental quality control regulations is illustrated by the increasing restriction on private incineration of mixed solid residuals. Many apartment house and office building incinerators have been shut down because they were unable to reduce their discharge of particulates to a level conforming to state or city air quality regulations. The buildings had been discharging a major portion of their solid residuals into the atmosphere, creating an external or social cost in the form of soot that dirtied the landscape and the lungs. The requirement that incineration not degrade the ambient air meant either that improved incineration had to take place, generally at greatly increased expense, or other means of disposal had to be found, also at increased expense.

The feasibility of recycling solid residuals originating in households, offices, and smaller commercial establishments depends largely on the unquantified costs and externalities of on-site handling, processing, and storage. A critical factor in these considerations is the extent to which the residuals can be kept separate and free from contamination. In large manufacturing and commercial establishments where all labor and equipment and materials are costed out, the exercise of quantification has already taken place, and decisions have been made on the most efficient use of materials.[2] But to date discussion of householder separation of residuals has generally been dismissed with the assumption that the householder is concerned solely with minimizing inconvenience.

The potential for recycling material may be enhanced or diminished in on-site handling, storage, and processing, or it may be completely destroyed. What happens is a function both of the material quality of the residual and the processes to which it is subjected. The storage of paper residuals mixed with other solid residuals severely diminishes the potential for recycling because, as explained elsewhere, the heterogeneity of the cellulose fibers recoverable from MSR limits the products that can be made and hence the markets for the recovered material. Materials such as ferrous metals, which are not modified by contact with other residuals and which are mechanically separable from them, may not lose their recycling potential if they are stored as mixed solid residuals. Of course, if a separated condition exists at time of generation and can be economically maintained, the additional inputs (i.e., energy) to achieve separation are not necessary.

[2] This assumes that externalities created by the discharge of liquid, gaseous, and solid residuals have been internalized by the imposition of environmental quality standards.

Increased density of material is as important for recycling as it is for discard. This is illustrated by Franklin and Darnay[3] who cite the difference in weight between a cubic yard of uncompacted steel and aluminum cans and a cubic yard of steel and aluminum cans which have been compacted to natural density—a fully densified state with all air space eliminated. The weight of the first is about 300 pounds, with 98 percent of the space taken up by air. The weight of the second is over 6 tons! Densification greatly reduces the space required for storage and transportation of material. In this respect on-site processing for recycling may be no more costly than on-site processing for final disposal.

Segregated residuals may take up more space in storage than mixed solid residuals simply because different locations are required for different categories of stored residuals. A prime example of inadequate space for the separate storage of residuals can be found in the federal government buildings in the District of Columbia. In the fiscal year ending June 30, 1970, the government sold about 21,000 tons of paper for recycling, the vast part of which was mixed paper residuals in bags and bales. Between 50,000 and 60,000 tons of paper were loaded out for incineration or landfill. The reason given for not recovering additional quantities of paper from the residuals of government offices is the shortage of storage space.

What space is available in locations suitable for storing residuals is usually preempted for parking. However, there may come a time when local governments will set mandatory space standards for on-site processing and storage of solid residuals, just as they have established requirements for parking space according to building use and size.

Collection and Transport

The second major element in a solid residuals handling and disposal system is collection and transport. Collection may be described in terms of the physical operations performed, the management agency performing the operations, the available technology and how it is used, and the relationship to the preceding and subsequent elements of the solid residuals handling and disposal system.

One of the most significant cost factors of a solid residuals collecting system concerns the relative locations of residuals generating sites. The

[3] Arsen Darnay and William E. Franklin, *The Role of Packaging in Solid Waste Management 1966–1976* (Washington, D.C.: U.S. Government Printing Office), p. 119.

horizontal density of the sites determines the relationship between the time spent in travel and the time spent in picking up residuals. Spatial arrangements at the site also determine the amount of time spent at each pickup stop. If municipal regulation provides for the final on-site storage point of the MSR at the back door, pickup time (or pickup labor) may be more than doubled.

The kind and amount of MSR collected depend on how much is generated, who does the collecting, how often they do it, and the characteristics of the urban environment in which the collection is made. A general description of the nature of mixed solid residuals found in municipalities was presented in chapter 2. Specific contents vary according to the climate of the municipality, the population density, income levels, cultural patterns of consumption, and the environmental quality controls that are practiced.

Collection practices affect the difference in quantity between solid residuals generated and collected. The American Public Works Association asserts[4] that total refuse quantities increase 40 percent when a route is collected twice a week rather than once a week. This datum is based on a survey made of several Chicago routes in 1967. Presumably the 40 percent not picked up in the once-a-week collection is picked up partly by street cleaning, partly by catchment basins, and is partly redistributed by the wind. Some portion is undoubtedly added to the neighborhood inventory of mixed solid residuals that slowly degrades to natural elements. Some part of the decline in weight of the once-weekly collection may also be due to evaporation of the moisture present at time of generation.

The collecting arrangements for municipal mixed solid residuals have significant ramifications for costs and for recycling. These arrangements vary from individual consumer hauling in suburban or rural areas to city contracts with one or two collectors who pick up all residential and commercial MSR in very large urban areas.

Probably the most common pattern is one with a variety of collecting agencies. Municipally operated departments collect from residential housing of three or four dwelling units or less. MSR from garden apartments, high rises, and commercial and manufacturing enterprises are collected by private carters working on contract or in some cases by the housing management. There are well over 100 private haulers in metropolitan Washington.

[4] *Refuse Collection Practice,* 3rd ed. (American Public Works Association), p. 475.

This diffusion of the collection function increases the congestion of city streets and contributes to the increasing costs of disposal because of the congestion that occurs at disposal sites. The cleanliness of the urban environment is also affected. There are increased opportunities for refuse to blow from trucks, and when a private hauler fails to collect from the building of a landlord who is delinquent in the payment of his fees, trash piles up.

The kind of collection equipment used is intimately related to on-site handling, storage, and processing activities and also to the arrangement of the building where generation occurs. Possible future systems include piped slurry and pneumatic tube installations for ground-up refuse. However, the procedure that promises to remain dominant for the foreseeable future is collection by truck. The biggest change that has taken place in the collection truck is the application of hydraulics to compress the loaded residuals at about a 3 to 1 ratio. The compactor truck is easier to load from street level than the high-sided truck in general use until fairly recently. A common procedure was for some of the crew to dump refuse containers into burlap bags which were then slung up to the man riding the load, who emptied them. At present the containers are tipped directly into a rear hopper which lifts the MSR hydraulically into the charging chamber of the truck body. The increased collection efficiency of the compactor truck is matched by the increased noise of its operation.

At sites where it is possible to use special hoppers or bins on casters for residuals storage, pickup may be made by a front-end loader which lifts the hopper up over the truck cab and tips it into the charging chamber. Front-end loading of residuals can reduce crew size by 50 percent or more. It is also leading to a new feature of the cityscape which some people may find objectionable—the presence of these bins on sidewalks by the side entrances of stores, or in parking areas of garden apartment complexes. In addition to rear-end loaders, side loaders, and front-end loaders, experimentation is going on with a collection truck equipped with a long hydraulic arm which can lift an MSR container from the curb and over a parked car, dump it in the truck, and return it to the curb.

The collection technique must be integrated with on-site operations in the solid residuals handling and disposal system. In a system in which residuals are baled on-site for transportation, there is no need for a compacting mechanism on the collection truck. There is equally strong interdependence between the mode of collection and the processing–

32

transfer element shown in figure 2. A collection truck equipped with a compacting mechanism and carrying a crew of three or four men for rear-end loading is an expensive vehicle for a 10-mile or longer trip to a landfill. Not only is the compacting equipment costly, but it takes up capacity which could otherwise be utilized for payload. Consequently, the use of compactor-equipped trucks tends to increase the use of transfer stations where MSR are transferred to hauling equipment which can carry larger payloads unburdened with compaction equipment and collection crews.

The cost of collection has been commonly cited as the major solid waste problem of municipalities, with 80 or 85 percent of residuals management costs chargeable to collection and 15 or 20 percent chargeable to disposal. This view is an oversimplified one derived from the expenditure records of municipal sanitation departments. Missing is the cost allocated to cover the initial activities of the system—on-site handling, processing, and storage. Also, this figure reflects a disposal cost based primarily on the use of the town dump before the imposition of environmental quality standards increased the costs of final disposal. This figure does not include in the cost of disposal the opportunity costs of the land used for the dump. Furthermore, that portion of transportation following last pickup, appropriately chargeable to disposal, has in the past been tallied as a collection cost. The exclusion of the cost of on-site operations from total system cost in most thinking outside of the self-contained system of the industrial plant has tended to obscure the reciprocal cost relationship between on-site operations and collection. Residuals packaged for ease and speed of loading are less costly to collect than residuals set out in random fashion.

The proper identification of all of the costs of the system, the division of management of the system among different agencies, and recognition of differences of scale in the generation of residuals are important elements in the consideration of externalities and incentives for a more efficient system. The components of collection cost for a small generator are markedly different from those of a large generator, with a resultant difference in potential incentives.

It would be very difficult to design a financial incentive that would reduce the residuals output of an individual household generator. Annual per capita generation of municipal residuals is about 1 ton collected over the course of 100 or so different pickups. Average cost of collection of 1 ton of mixed solid residuals is $15 to $20. The major component of this cost, however, is not the quantity of residuals collected. Clark *et al.* have reported[5] that collection frequency, pickup location (curb or rear yard), pickup density (number of residential pickups per square mile), and the method of financing collection were the most significant of nine factors examined in the cost of picking up MSR from a single residential unit. Assessing a household generator of residuals for the small portion of the cost of collection attributable to the weight or volume of residuals collected from his generation would not appear to provide a practical incentive.

The situation of the large-scale commercial generator is quite different. The weight or volume of generation is the most significant index of cost, because there are no dispersed points of collection. In the case of the large-scale generator, a single management agency has control over not only the handling, processing, and storage of residuals at the site of generation, but also over the collection and disposal elements of the system. With single-agency management, the generator is in a position to immediately realize the benefits of any efficiencies in materials use that he establishes.

Collection of Residuals for Recycling

The extent to which residuals can be recycled into the product material from which they were generated is influenced by whether they have been separated or mixed in their progress through the on-site and collection segments of the solid residuals management system. On-site separation at time of generation is essential. The mixture of residuals causes both a loss of identity and a contamination of one material by another, giving rise to the general judgment that after mixed collection consumption residuals can be recycled only into product materials of a lower grade than the original product, if at all. Thus the basic question on the relationship of the solid residuals management system to recycling is how this system operates for separated residuals in both on-site handling and off-site collection.

The collection of separated solid residuals for recycling is carried out in a variety of ways. A traditional way is for the generator himself to take old newspapers or copper or scrap iron to a junk dealer for the few pennies he

[5]Robert M. Clark, Betty L. Grupenhoff, George A. Garland, and Albert J. Klee, "Cost of Residential Solid Waste Collection," *Proceedings of the American Society of Engineers*, 97 (SA-5): 563–568.

will get for it. Another procedure involves the so-called gypsies, who, particularly when secondary materials markets rise, collect scrap and sell it to a junk dealer. A widespread practice is for Boy Scouts and church groups to stage newspaper collection drives several times a year. In other instances a secondary materials dealer himself will collect residuals from generating sources. Such collection generally is confined to industrial and commercial generators, though in some instances paper stock dealers collect old newspapers from apartment houses and churches. Recent citizen interest has stimulated separate collection of newspapers by municipal sanitation departments in several localities. This operation will be examined in more detail in chapter 5.

The economics of collecting residuals for recycling depend to a great extent on on-site processing operations. For example, used corrugated containers must be bundled or baled by the generating source in order to be picked up for recycling. The minimum on-site processing required in many instances is segregated storage to reduce contamination, and baling to facilitate the mechanical handling of the residuals. Compactor trucks are not needed for collecting baled residuals.

The collection of separated solid residuals is frequently cited as being too costly to permit recycling to be an economic activity except where collection is subsidized, as it is when it is performed by social groups or someone who is not otherwise employed. There are two fallacies with this kind of reasoning. First, the solid residuals are going to have to be collected either for landfilling or recycling, so it is the relative cost of separated collection for recycling that is significant in the total picture. It is the difference in cost between collection as a separated residual for recycling and a mixed solid residual for disposal which should be one of the factors influencing the economics of recycling.

Second, while the donated or less expensive labor of the Boy Scout or the unemployed gypsy gives the appearance of low cost or no cost collection of residuals aggregated for recycling, there are a variety of unpriced costs involved, including the contribution of the collection vehicle to traffic and air pollution. This is also true of recycling centers. The delivery of small quantities of residuals by many individuals is less efficient in conserving both energy and the environment, not to mention labor, than collection by a single agent. Individual trips are economic when separated residuals are returned to stores at the same time a trip is being made for other purposes—such as returning bottles on a grocery shopping trip.

Processing and Transfer

The next element of a solid residuals handling and disposal system following collection is processing and transfer (figure 2). These are distinct activities which may be combined at this point in the system, or processing may precede final disposal. A simple system may not include either activity.

A transfer operation is placed in a system for two reasons. As pointed out earlier, it is not efficient to use collection labor and equipment for a road haul of residuals beyond a 10- mile[6] radius from last point of pickup. Hence one or more transfer stations may be used to load the residuals on truck or rail for transport to more distant points.

Processing operations are generally designed to reduce the volume of solid residuals prior to final disposal. The most common form of processing over the years has been burning. Depending on the techniques used, burning converts a certain amount of the carbon–hydrogen molecules in solid residuals to gaseous residuals of carbon dioxide and water vapor, and deposits some solid residuals in the air in the form of particulates. Incineration emits to the atmosphere 10–60 pounds of particulates per ton of residuals burned, depending on the efficiency of combustion and the controls on the incinerator. Incineration may reduce volume of solid residuals by as much as 90 percent and weight by about 50 percent.[7] Incinerated residue does not need to be compacted for landfill because burning has reduced it to an average density of 1,000 pounds per cubic yard at time of landfill weigh-in, following water quenching at the incinerator.

A second method of reducing volume is grinding. This process produces mixed solid residuals of a small and uniform particle size, reduces air spaces, and gives them a density six to seven times greater than loose MSR. Because ground residuals have an increased surface area exposed to oxygen, they decompose more rapidly than residuals discarded without grind-

---

[6] R. J. W. Devanney, Vassilios Livanos, and James Patell, *Economic Aspects of Solid Waste Disposal at Sea*, A report prepared for the National Council of Marine Resources and Engineering Development (Cambridge, Mass.: Massachusetts Institute of Technology Press, 1970).

[7] Jack DeMarco, Daniel J. Keller, Jerold Leckman, and James L. Newton, *Incinerator Guidelines, 1969* (U.S. Dept. of Health, Education and Welfare, 1969) claims a total reduction of 75–80 percent of weight of the as-charged solid waste, including moisture by incineration to a (bone-) dry residue (page 1). The figure of 50 percent is used here as being more representative of common practice and more nearly reflecting the state of the residue at time of landfill.

36

ing. Biological studies of ground MSR in landfills without cover have shown them to be inhospitable to rats and not productive of flies.[8] Ground-up refuse can be transported at relatively high densities.

Other processing operations at stage 3 in figure 2 may be compacting and loading in a high capacity (70 cubic yards) over-the-road trailer body, and baling. Baling may be particularly appropriate for rail haul and, like grinding, eliminates the need for compaction in the course of the landfill operation.

The measurement and allocation of processing and transfer costs in a solid residuals handling and disposal system have not reached the stage of an accepted accounting convention. As mentioned above in the discussion of collection, the oft-repeated statement that "collection accounts for 80 percent of waste disposal cost and landfill 20 percent" considers the transportation from the last pickup point to landfill as a component of the collection cost. The increasing costs of transporting MSR greater distances to available land and the advent of the transfer station to increase system efficiency are bringing about a realization that the cost of transport from last point of pickup is appropriately ascribed to disposal rather than collection cost.[9] A cost-accounting scheme prepared for private contractors by a public accounting firm allocates 20 percent of route labor costs and 20 percent of collection equipment operating costs to disposal.[10] It was the judgment of the accountants that generally one fifth of the time of a collection crew and equipment is spent on disposal. This is an important datum in considering the economic feasibility of recycling, because, while collection cost is appropriately assessed for residuals to be recycled as well as for residuals to be disposed of, residuals going to recycling do not incur the costs of disposal which should (but frequently do not) include the costs of final hauling as well as the costs of incineration and landfill.

Mechanical separating devices used at transfer and processing stations to recover recyclable fractions of MSR are classified as dry systems and wet systems. Dry systems may rely on the aerodynamic property of the

[8] Robert K. Ham, Warren K. Porter, and John J. Reinhardt, "Refuse Milling for Landfill Disposal," *Public Works* (January 1972), p. 72.
[9] An accounting system which treated the haul to disposal as a transportation cost would be equally acceptable so long as it was differentiated from the cost of collection.
[10] *1966–67 Sanitation Industry Yearbook* (New York: Communication Channels, Inc.), p. 20ff.

materials, which are separated in columns of air; they may utilize light reaction to a photoelectric cell or simply magnetism. The only dry separation systems in fairly widespread use at present are those used in magnetic extraction of ferrous metals. A combined manual—mechanical method of separation involving manual removal of MSR fractions from moving conveyor belts is in use in some cities of the world. It is dependent on relatively low wage scales. Wet separation is carried out by putting the MSR in flowing water and running the mass through a configuration of screens and centrifuges. A pilot plant utilizing wet separation is operating in Franklin, Ohio. It has been successful in separating marketable fractions of glass and metals from mixed solid residuals. It has also been successful in reclaiming a portion of the cellulosic fiber in the mixed solid residuals, but in such a condition that they are usable only as a component of roofing felt. The separation of recyclable material from mixed solid residuals appears to be feasible only for materials not seriously contaminated by mixture with other materials, and having physical characteristics that facilitate mechanical separation, such as the magnetic properties of ferrous metals and the density of glass. There is substantial doubt whether postcollection separation can yield materials for recycling except at a lower grade than the original product material. It is not yet clear under what conditions such operations are economical. Wet or dry separation at stage 3 requires additional inputs of material and energy to accomplish the separating, thereby generating more gaseous and other residuals in the process.

There are other types of recovery operations which can take place at this stage of the solid residuals handling process. The Bureau of Mines has done extensive research on recovery of recyclable fractions from incinerator residue. Several groups, including the Bureau of Mines, are working on pyrolysis, a high-heat, oxygen-free process that can turn organic materials into oils, tars, and recoverable gases. A number of studies have been made of the feasibility of composting MSR for a soil conditioner. There are also several steam plants in operation which use mixed solid residuals as a fuel. The economies of these operations have not been finally proven in this country.

The on-site storage and processing stage (1 in figure 2) and the processing and transfer stage (3) are the two places in the solid residuals handling and disposal system where residuals may be diverted from the stream flow to disposal. It is clear that residuals diverted at 1 can be used to produce the same material that comprised the original product. This is much less possible at box 3 after mixed collection. Total enumeration of the costs in the varying circumstances has yet to be worked out.

Landfilling

The final element of the solid residuals management system is the sanitary landfill. Sanitary landfill is sometimes characterized by cynics as a euphemism for a dump. Dumps have existed from the time human beings congregated in villages. They generally are characterized by ugly appearance, bad smells, smoldering fires, flies, rats, blowing paper and other debris, and are often located downwind of the more affluent neighborhoods. They are infrequently covered over with dirt, if at all, and often no consideration has been given to their potential contamination of groundwater. Dumps have been partly out-of-sight, out-of-mind depositories viewed as useful by the refuse hauler, the archeologist, and the antique hunter, and, insofar as possible, ignored by the rest of society.

A sanitary landfill is quite different from a dump. First, a good landfill is properly sited. Consideration has been given not only to alternative uses of the land but to the ecological effect that landfilling will have on a particular site. Marshes and riparian lands, once considered good places for dumps because they were considered poor for building, are now recognized as some of the most biologically productive areas that exist, and are no longer automatically selected for disposal sites by solid waste management personnel. Sites may be ruled out for landfills because of the possible contamination of ground water by the leachate produced by the landfill; or, to prevent contamination of groundwater, the landfill area may be lined and drainage provided to direct the leachate through a sewerage treatment process. Second, the ultimate use of a possible landfill site is considered. It is generally slated to become a recreational area. A landfill area is not considered a good area for heavy construction because continuing chemical and biological activity in the deposited residuals plus natural subsidence make the land somewhat unstable for a period of years. A well-compacted landfill can be used for some forms of construction in 4 or 5 years.

There is reference these days to "the coming crisis in refuse disposal" because, some say, we are running out of land. Skeptics wonder how this can occur in the continental United States, which has a population density of one person for every 11 acres; surely there is room to deposit on these acres the 6 or 7 pounds of municipal solid residuals generated per capita every day. The answer, of course, is that there appears to be a shortage of *suitable* land *available* for sanitary landfills. The suitability of land refers to distance of the site from generating source, alternative uses, ecological compatibility, ground water orientation, and future use of the completed landfill. Suitability includes consideration of access routes. Landfill sites 300 miles or more distant from the generation area are being considered today.

The availability of sites is as limiting a factor, if not more limiting, than considerations of suitability. The Office of Solid Waste Management of the Environmental Protection Agency has estimated that, on the average, existing landfill sites have capacity for only another 4 or 5 years under assumptions of the achieved density of MSR and the final contours of the fill. Ninety-four percent of the 6,000 land disposal sites sampled by the Federal Solid Wastes Management Office did not meet the most modest criteria defining a sanitary landfill. These data are reflected in newspaper stories citing the plight of one locality or another which has been denied access to an area for landfill use by political or court action started by citizens who do not want such operations in their areas. Thus, availability may be a matter of economic distance, appropriate land use, or political consensus. Adverse political reaction in many instances reflects previous bad experience with dumps.

Site design and preparation of a landfill area involve landscaping, fencing, construction of all-weather access roads to dumping areas, construction of the weigh scale area, and provisions for prevention of groundwater contamination by leaching. Operations include weighing in and tallying loads for record and billing purposes, preparing cells, compacting dumped MSR, and covering with fill. If the system does not provide for processing operations, such as grinding, at stage 3, they may take place here.

The amount of land required for a landfill depends on a number of factors, including the compactibility of the residuals, the degree of compaction actually applied, the design height of the fill, and the desired life of the landfill. Mixed solid residuals arriving at a landfill in a loose state weigh 150 to 180 pounds per cubic yard. Compactor loads have a density of 500 to 700 pounds per cubic yard. Compacted refuse as it finally reposes in the landfill prior to decomposition, but after the application of the dirt cover, is generally considered to weigh about 1,000 pounds per cubic yard. Ralph Stone[11] calculated densities of 852 to 1,250 pounds per

[11] Ralph Stone, "Refuse Landfill Compaction with Crawler and Wheel Type Equip-

cubic yard, dry weight, for landfill cells in an experiment testing the efficiency of various kinds of compacting tractors and crawlers. Incinerator residue, like demolition residuals, arrives at a landfill uncompacted at about the same density it will achieve with compaction, 1,000 pounds per cubic yard or more, while compacted corrugated container residuals may arrive at a landfill weighing less than 100 pounds per cubic yard.

The operation of containing mixed solid residuals in a landfill so that they will not blow about, burn, or serve as hosts for pests requires covering and compacting with earth fill following compaction of the solid residuals. A minimum requirement is placing 6 inches of compacted earth fill on the mixed solid residuals at least once a day, and a compacted layer of earth 2 feet deep on the finished grade of the landfill. The ratio of fill to mixed solid residuals is specified at about 1 to 4. Five yards of landfill capacity are required for every 4 cubic yards of solid residuals, the fifth yard being for dirt fill. The significance of the compacted earth cover is evident from the fact that "house flies emerging from the pupate stage can crawl up through more than five feet of loose soil, but they cannot penetrate through six inches of compacted soil."[12]

The commonly accepted costs of landfilling have, in the past, been among the least accurately assessed costs of the solid residuals handling system. This has been due to the externalities of final disposal. Externalities include the deleterious effect of high-volume truck traffic on the area adjacent to access routes, including the polluting effect of the exhaust emissions; blowing paper or other dirt from landfilling operations; fermentation gases which offend the nostrils; and the generally unpleasant appearance of the site. While these costs are not borne by the system, they may be measured in part by the extent to which these damages depress the value of adjoining land.

When the cost of landfilling is fully internalized, the accounting cost in the system obviously is considerably greater; however, adjoining land values do not suffer. By investing in adequate design and engineering operations and by adopting adequate operating techniques, the impact on adjacent property can be minimized. Wide and screened access rights of way can shield landfill traffic, which can also be reduced by load con-

---

ment," paper presented at Engineering Foundation Research Conference, Beaver Dam, Wisconsin, July, 1968.

[12] *California Solid Waste Management Study 1968 and Plan 1970,* U.S. Environmental Protection Agency (Washington, D.C.: U.S. Government Printing Office, 1971), pp. vii–5.

solidation at transfer points. Fencing and natural screens can mitigate the unpleasant sight of landfill operations. Controlled unloading and compacting operations can eliminate the blowing of paper and dust. These elements will place the cost of landfilling where it properly belongs, on the operation itself. In some areas, Los Angeles for one, adjacent land values have actually increased because the landfills have been well designed and well operated and the completed fill has become a desirable recreation area.

4

Flow of Paper Through the Production-Consumption Sequence

Basic Varieties of Paper

The word *paper*, like the word *metal*, is a general term designating a wide variety of substances made up for the most part of cellulosic fibers. One hundred years ago most of the paper in the United States was made from used cotton cloth; today 98 percent of the cellulosic fiber in paper comes from trees, the other 2 percent coming from cotton, flax, straw, and bagasse (the residue of sugar cane).

There are four basic ways of refining the cellulosic fiber in pulpwood for paper. The technology that is applied depends on the kind of wood used and the qualities, such as strength or brightness, that are required in the final product. The most widespread process is the kraft chemical or sulfate process which produces paper that has long fibers and considerable structural strength. The kraft process is used to make paper for grocery bags and corrugated shipping containers, among other products. Of the 44 million tons of wood pulp consumed in 1969, 30.6 million tons were varieties of sulfate pulp.

43

A second chemical pulping process, and an older one, is the sulfite process, which is primarily limited to pulpwoods with little resin content. The chief advantage of the sulfite process is its ability to produce a relatively bright pulp without bleaching and a pulp easily bleached to higher brightness. This process is used for fine papers for stationery and printing. Consumption in 1969 was 2.7 million tons.

A third method of preparing cellulosic fibers for making paper is a semichemical pulping process referred to as NSSC, an acronym for neutral sulfite semichemical pulping. This process includes chemical cooking to start the separation of fibers, a process that is completed mechanically. A major use for semichemical pulp is corrugating medium, the center ply of double wall corrugated fiberboard. Semichem corrugating medium has a stiffness that gives columnar strength to the flutes of the corrugated board. Consumption of semichemical pulp in 1969 was 3.6 million tons.

The fourth major category of wood pulp is groundwood, which accounted for 4.7 million tons of domestic consumption in 1969. Groundwood pulp is made by a mechanical process utilizing stone or mechanical grinders. In the chemical processes, lignin, which is nature's glue holding the cellulose fibers in a tree together, is dissolved partially or entirely out of the wood, leaving discrete cellulosic fibers to be reformed into a sheet of paper on the paper machine. In groundwood processes the lignin remains a part of the separated fiber. The major use of groundwood pulp is the manufacture of newsprint.

Paper Products

A great variety of papers are manufactured from these four kinds of pulp. The list[1] starts with abrasive papers (heavy kraft) and adding machine paper and includes automobile board in the A's. The B's include barber's headrest paper, bedstead wrapping paper, and bill straps (used to bundle currency). Bristols are called Bristols because this kind of paperboard was originally made in Bristol, England. The C's include casket paper (basis strength 50 pounds) used to fluff up the cloth lining of a casket, and cup board, a paper for making paper cups. The major E is envelope paper. Manila paper is so named to indicate the color and finish that used to be obtained with paper manufactured from manila hemp

---

[1] *The Dictionary of Paper*, 3rd ed. (New York: American Paper and Pulp Association, 1965).

stock. The list ends with Yoshino—a Japanese tissue paper made from the fibers of the paper mulberry.

Classification of different kinds of paper is not very orderly because of exceptions to the categories and the ambiguities of terms. The generic term *paper* includes both paper and paperboard. Generally, if paper is more than twelve thousandths of an inch thick (termed "12 points"), it is classified as paperboard. However, blotting paper and some drawing paper thicker than twelve thousandths are designated as paper, not paperboard. Governmental and industrial statistical reporting follows this division. Of the 59 million tons of paper consumed in 1969, 30 million tons were classified as paper, and 28.8 million tons were classified as paperboard.

Classification within these categories tends to follow the lines of paper manufacturing—the kind of furnish used, for example sulfite, sulfate, or reclaimed paper pulps; bleached or unbleached; the nature of the finish, for example, coated or uncoated; and end use, for example newsprint or book paper. In some cases the manufacturing process alone is the determinant of the category. For example, wet machine board is placed in a separate category though its annual tonnage is small (162,000 tons) and its uses as hard book covers and shoe board are highly specialized. Table 2 lists the major categories of paper and paperboard consumption in 1969. The consumption figure is derived from production plus imports less exports.

Almost 50 percent of the paper and paperboard used in 1969 was converted into packaging products. It is estimated[2] that 25 percent of paper (about 7.5 million tons) and 85 percent of nonconstruction paperboard (about 20.4 million tons) were used to make the containers, wrappings, and bags that package goods produced in the United States. Of the 58.9 million tons of paper and paperboard consumed, 28 million tons were used for packaging.

The largest component of this tonnage, about 14 million tons, or over half of all packaging paper and board, was used to fabricate corrugated shipping containers. These are used principally for the distribution of food and kindred products, 27.6 percent; paper and allied products, 11.3 percent; and glass products, 10.2 percent.[3] Other components are folding

[2] Arsen Darnay and William E. Franklin, *The Role of Packaging in Solid Waste Management 1966–1976* (Washington, D.C.: U.S. Government Printing Office, 1969), p. 8.

[3] *Fibre Box Industry Statistics, 1969* (Chicago: Fibre Box Association, 1970), p. 42.

TABLE 2
Major Categories of Paper Consumption,
1969 (millions of tons)

Grade	Quantity
Paper	
Newsprint	9.9
Coated printing and converting	3.3
Book paper, uncoated	2.8
Writing and related papers	2.9
Unbleached kraft packaging and industrial converting	3.7
Tissue and other machine creped	3.6
Other	4.0
	30.2
Paperboard	
Unbleached kraft packaging and industrial converting	10.0
Bleached packaging and industrial converting	3.3
Combination furnish paperboard	7.3
Semichemical paperboard	3.5
Construction paper and board	4.6
	28.7
Total	58.9

Source: American Paper Institute, *The Statistics of Paper 1972.*
[a]Consumption equals production plus imports less exports.

boxes (also used primarily for food distribution), set-up paper boxes used primarily for wearing apparel, and paper sacks.

Major end use categories for nonpackaging paper in 1969 were newsprint, 9.9 million tons; printing papers, 6.5 million tons; and tissue paper, 3.5 million tons.[4] More than 90 percent of newsprint is used in newspapers. Printing paper is used for books, magazines, and commercial printing as well as envelopes, writing tablets, and adding machine tape. The third largest group, sanitary tissue, includes toweling, toilet tissue, table napkins, and facial tissue.

[4] Newsprint and tissue paper amounts are taken directly from *The Statistics of Paper* (New York: American Paper Institute). The quantity of printing paper is estimated on the basis of 1966 statistics given in Darnay and Franklin, *The Role of Nonpackaging Paper in Solid Waste Management,* p. 9.

These are the kinds and quantities of paper and paperboard used by our society. The two kinds of paper selected for intensive investigation in this study—newsprint, the major end use of paper, and corrugated container board, the major end use of paperboard—together account for over 40 percent of total paper consumption.

Newsprint and Newspapers: The Products

Newsprint is made from a combination of groundwood pulp and chemical pulp formed on a fourdrinier paper machine. Standard newsprint, as defined by the United States Customs Court for tariff purposes,[5] is paper having a basis weight of 30 to 35 pounds (the weight of 500 sheets measuring 24 by 36 inches), a machine finish with a gloss of not over 50 percent by the Ingersol Glarimeter test, a maximum caliper of 0.0042 inches, and a maximum ash content of 6 1/2 percent. The common basis weight of 32 pounds works out to a weight of 1 pound for an 80-page edition of the daily *New York Times*. Important characteristics of newsprint include pressroom runability (measured as breaks per 100 rolls of newsprint run on the presses), printability, opacity, tear strength, and brightness.

Two product characteristics of newspapers that are of interest in this discussion are the dimensions of the newspaper page and the thickness of the published newspaper. These two characteristics, along with number of copies produced, determine the amount of newsprint residuals to be disposed of in the production–consumption system. The thickness of a newspaper is primarily a function of the number of pages, and is only slightly affected by the caliper of the pages.

The dimensions of a newspaper page were established over 260 years ago in 1711 when the British government, in an effort to suppress "increasingly impudent" newspapers, levied a tax on the basis of each page printed.[6] The reaction to this government fiscal incentive was indeed fewer, but much larger, pages. The establishment of 720 square inches of printed surface in one plane has contributed to a product shape which has profound implications for recycling, as discussed later in this chapter.

The number of pages in the average newspaper edition with a circulation of 50,000 or more has more than doubled since World War II, from 23

[5] *Dictionary of Paper*, pp. 309–310.
[6] Ben H. Bagdikian, *The Information Machines: Their Impact on Men and the Media* (New York: Harper and Row, 1971), p. 224.

pages in 1946 to 47 pages in 1969, an increase of 104 percent.[7] The average number of pages devoted to news, editorial, and other nonadvertising content rose from 10.4 in 1946 to 17.8 in 1969, an increase of 71 percent. In 1969 paid advertising space occupied an average of 62 percent of the average newspaper's pages as against 55 percent in 1949. Paid advertising space expanded 132 percent during this period. These statistics cover only the lineage that is printed in the newspaper publishing plant. Inserts printed in outside plants but distributed with newspapers, particularly on Sunday, are not included. If they were, the proportion of space devoted to advertising would be substantially higher.

Sunday papers in the United States are a particularly striking example of increased consumption of newsprint in recent years. In some cases they equal or exceed the weight of the total issues for the other six days. The championship status of the newsprint-heavy *New York Times* is acknowledged in the following verse written by Larry Hirsch of the Norfolk *Virginia-Pilot*:[8]

> I think that I shall never see
> A *Times* much thinner than a tree
> That is, a *Times* by presses born
> Prodigiously for Sunday morn.
> A weekday's *Times*, though also big,
> Compared to Sunday's is a twig,
> So weighty is the news that's fit,
> You risk a rupture lifting it.
> So thick is it, though it's not rolled,
> It takes at least two hands to hold.
> And were it rolled—a trunk of dead wood—
> It's rings of growth would rival redwood.
> O great green groves of Kilmer's muse
> Converted into great gray news,
> O trees that fall to axes' strokes—
> Not one is mighty as the Ochs.

[7] Jon G. Udell, "Economic Trends in the Daily Newspaper Business, 1946–1970," Graduate School of Business, University of Wisconsin, Madison, December 1970. (A research publication of The Bureau of Business Research and Service.)
[8] *Trends of the Times*, June 1969. A brochure issued occasionally by the promotion department of the *New York Times*.

In taking notice of this verse, the *Times* noted that it uses more newsprint than all the newspapers of any *state* except California, Illinois, Ohio, Pennsylvania, and New York. Their consumption of newsprint in 1969 was 400,000 tons or about 4 percent of total newsprint consumption in the United States.

The number of pages published by one newspaper in a week's time is intimately related to the amount of advertising carried by a newspaper. In turn, the amount of advertising is intimately related to the density of population, a high population area having more lines of advertising. This has important ramifications for the generation of newspaper residuals because not only will more newsprint be consumed by the higher number of newspaper readers in a densely populated area, but also the newspapers consumed will be thicker because of the increased advertising. It is also reasonable to expect that areas of higher median income will have more newspaper advertising than areas with the same population density but lower median income.

Of the 9.9 million tons of newsprint consumed in the United States in 1969, 6.8 million tons or 69 percent was imported, and the balance of 3.1 million tons was produced by U.S. mills.[9]

The growth in newsprint consumption from 1946 to 1969 has about paralleled the increase in gross national product from 4.3 million tons in 1946 to over 9.7 million tons in 1969. Based on assumptions about future economic growth, circulation, and advertising lineage, it has been estimated that newsprint consumption in 1980 will reach 13.1 million tons.[10]

Corrugated Containers: The Products

As noted previously, corrugated containers and newspapers comprise a substantial proportion of the mixed solid residuals to be disposed of by municipalities. Newspapers become consumption residuals primarily in households. The prime function of corrugated containers is the protection of goods in transit and in storage. Relatively few corrugated containers become residuals in households; their major points of generation as residuals are in storerooms and warehouses where wholesale lots of goods are broken down for retail sale, in assembly plants, and in large office complexes receiving quantities of supplies in case lots.

[9] *The Statistics of Paper* (New York: American Paper Institute, 1971).
[10] J. G. Udell, *Future Newsprint Demand 1970–1980*, a study prepared for the American Newspaper Publishers Association, January 1971, p. 40.

The hogsheads of a hundred years ago and the wire-bound wooden boxes and crates and bushel baskets of more recent vintage have yielded to the corrugated shipping container as the principal device for protecting goods in transit and storage. Heavy tri-wall containers are used by movers for packing and transporting furniture that used to be crated. Corrugated containers protect shipments of fruits and vegetables from California and Texas to the East Coast.

The use of colloquial and inexact terms in the paper industry, reflecting the historical development of the craft, is very apparent in the shipping container board section of the trade. Both *container* and *shipping* are important words, distinguishing the product under discussion from *box* and *carton*. They denote packaging used for transit rather than simply for protection and display. *Box* and *carton* denote paperboard used for the individual packaging of retail goods. Suits and cereals come in folding boxes; perfume is often packaged in a set-up box; cigarettes are sold by the carton; and all of them are transported in shipping containers. It is not unusual for a layman to refer to all of these products as cardboard. The situation is further confused by grouping together corrugated fiberboard and solid fiberboard in statistical tables; both use the same pulp furnish for the manufacture of shipping containers, but end up with structurally different materials. The final confusion, which contradicts some of the above, is that the manufacturer's organization concerned with the production of shipping containers is called the Fibre Box Association.

Fabrication of Corrugated Containers

Corrugated container board is manufactured in a variety of configurations of facings and walls. The characteristic structural element is the fluted or columnar element (corrugating medium) which is glued to a smooth sheet of paperboard called linerboard. When corrugating medium is glued to linerboard on only one side, the container board is termed single face; on two sides, double face. Double face corrugated paperboard is designated single wall. A combination of a liner, corrugating medium, liner, corrugating medium, liner is designated double wall, and the addition of one more layer of corrugating medium with a liner results in tri-wall. Linerboard has a caliper ranging from 0.008 to 0.028 inches; and a basis weight of 26 to 90 pounds. Corrugating medium ranges in caliper from 0.009 to 0.012 inches and in basis weight from 26 to 36 pounds. Usually basis weight is the weight of 500 sheets of paper or paperboard of a specific dimension. The basis weights of corrugating medium and linerboard, however, are given in terms of pounds per thousand square feet of the material.

Linerboard, the mill product that is made into the facing of corrugated container board, is made primarily from kraft pulp on a fourdrinier paper machine. Linerboard is also made on a cylinder machine from reclaimed kraft fibers and is referred to in the trade as "jute," a misnomer because no jute fibers are involved. Linerboard that is made from at least 85 percent new kraft fiber is designated virgin kraft in the trade, even though up to 15 percent may be recycled fiber. Alternatively, linerboard may be made from 100 percent recycled fiber.

Corrugating medium can be manufactured from a neutral sulfite semi-chemical hardwood pulp, from recycled fibers, and from wheat, oat, or rye straw. The major material is the semichemical pulp. The use of wheat, oat, or rye straw, widespread some years ago, is virtually unknown in the United States today.

It is in the manufacture of corrugated container board, more than any other segment of the paper industry, that one can find the explanation for the 50 percent decline in the rate of recycling for all paper residuals between 1944 and 1969. Of 2,260 pounds of fibrous materials used to manufacture 1 ton of container board in 1943, 900 pounds were recycled fibers and 920 pounds were kraft fibers, with 410 pounds of other fibers.[11] In 1963, 2,200 pounds of fibrous materials were used to make 1 ton of container board; of this amount kraft fibers comprised 1,360 pounds, semichemical pulp, which was not used in 1943–44, accounted for 380 pounds, while paper residuals were down to 420 pounds, a reduction of over 54 percent from 1943. This decline has continued to 1969. Since the production of corrugated container board now exceeds 14 million tons a year, the decline in percentage of paper residuals used in this segment of the paperboard industry is a very significant portion of the overall percentage decline. The reasons for this decline are due in part to improved kraft pulping technology, in part to the introduction of the new semichemical pulp, and in part to other factors.

Functions of Corrugated Container Board

The performance characteristics sought in the manufacture of linerboard and corrugating medium for shipping containers are more complex than

[11] Dwight Hair, *Use of Regression Equations for Projecting Trends in Demand for Paper and Board*, Forest Resource Report No. 18 (U.S. Department of Agriculture, December 1967). Adapted from table 13, p. 156.

one might expect. Both vertical and horizontal compression are important for the protective function of the container. Recycled container board made on a cylinder machine contains more fibers oriented in the machine direction than fourdrinier linerboard. The high ratio of machine direction fibers in cylinder board makes for greater stiffness in end-to-end compression. Recycled fibers also make a liner with a smoother and therefore better printing surface. It is also a better surface for sealing purposes, though both kraft and recycled container board may receive a variety of surface finishes in the papermaking process. Kraft liner is said to retain its strength across score lines better than recycled liner.

Corrugating medium is important to the structure of container board both for its columnar strength, which is generally exhibited in the vertical plane of the container, and for its resistance to crush, i.e., the flattening of flutes. Board that has poor resistance is said to have low flat crush. There are four different standardized flute sizes, A, B, C, and E, ranging in size from 36 flutes to the lineal foot for A to 96 flutes per lineal foot for E. A flute is considered to have the greatest compression strength, top to bottom, and the lowest flat crush. C flute with 39–42 flutes per lineal foot is the most commonly used design.[12]

The corrugated shipping container, unique among paper and paperboard products, has a legal status. The specifications for its design and construction are part of the contract (included in the small print on the bill of lading) between the shipper and the carrier (rail, truck, airplane, ship). If a product is damaged in transit, the liability may lie with the party who packaged the shipment if the packager did not do an adequate job; it may lie with the carrier for handling the shipment in a rough manner or exposing it to the elements; it may lie with the packaging itself for failing in a way it should not have. To assure that goods offered for shipment are adequately protected, the railroads in the *Uniform Freight Classification* (Rule 41) and the truck companies in the *National Motor Freight Classification* (Rule 5) have established certain specifications to be met by corrugated containers offered for shipment. Every corrugated container made has the box maker's certificate printed on it, attesting to the specifications that the container meets. If there is damage in a shipment, and the corrugated container on examination and test is found to meet the specifications set forth in the box maker's certificate, then the box maker has discharged his potential liability for the damage.

[12] Kenneth W. Britt, ed., *Handbook of Pulp and Paper Technology* (New York: Van Nostrand, Reinhold Co., 1970), p. 562.

A box maker's certificate includes the assertion that "this single wall box meets all construction requirements of applicable freight classification." For a given size and weight of contents these requirements might be bursting strength, 200 pounds per square inch; minimum combined weight of facings, 84 pounds per thousand square feet; size limit, 75 inches; gross weight limit, 65 pounds.

The key element in this specification is the bursting strength, otherwise known as the "Mullen," after the man who devised a test for corrugated container board in which a circular portion of it is subjected to pressure until it bursts. This test applied to liners of the same basis weight shows a greater resistance to bursting in liner made from virgin kraft fibers because they are longer than recycled fibers. Thus a 42-pound basis weight kraft liner will generally be able to sustain a bursting test of 200 pounds per square inch. To get the same bursting strength with recycled fiber, it is generally necessary for the liner to have a basis weight of 52 pounds per thousand square feet. Achieving this basis weight requires a greater caliper.

It is contended by some that Rule 41 is biased against the use of recycled fiber without justification because the prescription of a Mullen test is a material specification rather than a performance specification and may not accurately reflect the capability of a container to protect its contents from the normal hazards of shipment. Others say that Rule 41 does not serve as a container specification, but rather, by establishing grades combining bursting strength and basis weight, serves to standardize a highly variable material in such a way as to promote its wide acceptance in commerce over a large and climatically varied geographical area.[13] It is beyond the province of this discussion to examine the merits of this argument, but it is important to note its existence.

End Uses of Corrugated Container Board

The end uses to which corrugated container board was shipped from box plants in 1969 are shown in table 3. The greatest part, 27.6 percent, was used for packaging food and kindred products. The two largest categories within this group are meat products, 5.4 percent; and canning and preserving fruits, vegetables, and seafoods, 5.4 percent. Beverage industries accounted for 3.8 percent.

---

[13] George C. Maltenfort, "Freight Rule Requirements and Packaging Specifications—the Big Difference," reprinted from *Paperboard Packaging,* issues of April, May, June, August, September, October 1964, pp. 6, 7.

Recycling

TABLE 3
Classification of Shipments of Corrugated Container Board
by End Use, 1969

Major group[a]	Manufacturing industry	Percentage
20	Food and kindred products	27.6
21	Tobacco	0.6
22	Textile mill products	3.0
23	Apparel	1.4
24	Lumber and wood products	0.8
25	Furniture	3.5
26	Paper and allied products	11.3
27	Printing, publishing	1.4
28	Chemicals and allied products	4.3
29	Petroleum refining and related industries	0.9
30	Rubber and miscellaneous plastics products	3.7
31	Leather and leather products	0.6
32	Stone, clay, and glass products	10.2
33	Primary metal industries	1.2
34	Fabricated metal products	4.7
35	Machinery, except electrical	2.6
36	Electrical machinery equipment and supplies	7.3
37	Transportation equipment	3.4
38	Professional, scientific and controlling instruments, photographic goods	0.7
39	Miscellaneous manufacturing industries	9.0
90	Government	0.3
		100.0

Source: Fibre Box Industry Statistics, 1969.
[a]Standard Industrial Classification.

The significant component in the 10.2 percent shipped to the stone, clay, and glass products category was glass and glassware, which accounted for 7.9 percent. It should be noted that these statistics include corrugated containers shipped to manufacturers of glass jars. The statistics do not account for a container reused by a manufacturer of preserved fruits or by a bottling plant. For example, if a maker of mayonnaise reuses the corrugated container in which he received empty glass jars to ship full jars of mayonnaise to the grocer, the container is shown in table 3 as a container for glassware, not a container for canning and preserving products. To this extent the use of corrugated containers for food and kindred products is understated. The second end use category in size after food

54

and kindred products is paper and allied products at 11.3 percent. A major portion of these products consists of so-called disposables such as toilet paper, facial tissue, towels, and sanitary items.

The identification of the end use of corrugated containers gives some clues to the locations where goods are unpacked from the containers, and the containers become commercial consumption residuals. The largest category is supermarkets, the major dispensers of food and kindred products, paper and allied products, and the household supplies which come under the chemicals and allied products group. A smaller but still significant category is department stores. Automotive and electrical appliance assembly plants are a significant group. For example, the Dearborn assembly plant of the Ford Motor Company recovers over 100,000 tons a year of corrugated containers in which they have received parts for assembly into automobiles. Office buildings receiving shipments of paper goods and cleaning supplies have to dispose of substantial quantities of used corrugated containers. Comparatively few corrugated containers become consumption residuals in households for the simple reason that they are used, for the most part, for packing wholesale quantities of products which are sold at retail to individuals in smaller quantities.

Residuals of Paper Production and Consumption

A variety of residuals are generated in the production and consumption of paper products. This study is concerned only with the handling and disposal of paper residuals. Such residuals as the slash left in the forest from cutting pulpwood, the hydrogen sulfide emitted from the stacks of kraft pulp mills, or the suspended and dissolved solids discharged from a plant recycling newsprint are not considered.[14]

There are four stages in the production–consumption sequence of paper and paperboard at which paper residuals are generated as depicted in figure 3. The first stage is at the paper mill[15] where wood pulp, paper stock, and other fibers are manufactured into paper and paperboard. In 1969, 44 million tons of wood pulp, 12 million tons of waste paper, and a little less than 1 million tons of other fibrous materials were used in the manufac-

[14] Another RFF study by Bower, Löf, and Hearon is concerned with other residuals arising from the manufacture of paper.
[15] Many paper mills are integrated, that is, they produce both wood pulp and paper.

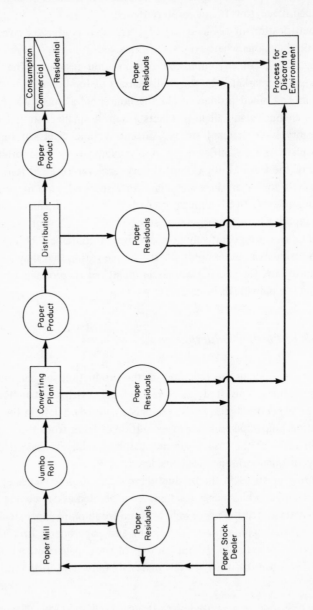

FIGURE 3

Stages of paper residuals generation and recovery of disposal.

ture of 54.1 million tons of paper and paperboard. The paper residuals from the paper mill operation, termed mill broke, are reused directly and are not a part of the solid residuals problem. Mill broke varies between 15 and 20 percent of the paper machine input,[16] primarily as a function of the type of product and machine design.

Converting Residuals

The second stage of the production–consumption sequence at which paper residuals are generated is in converting. It is at the converting stage that jumbo rolls of newsprint become newspapers; folding boxboard becomes suit boxes or cracker boxes; linerboard and corrugating medium are glued together and formed into corrugated container board for shipping cases.

The paper residuals generated in converting range from about 1.5 percent to as high as 20 percent of product input to converting. For example, in 1969 the folding carton industry generated over 600,000 tons of paper residuals in the manufacture of 2.6 million tons of products. In contrast, converting residuals generated in printing newspapers were between 1.5 to 3 percent, or about 250,000 tons in 1969. With total consumption of 58.9 million tons of paper and board in 1969, converting residuals probably amounted to about 5.5 million tons, of which 4.5 million tons were recycled.

Manufacturers of paper and paperboard products may encompass both mill operations and converting operations within the same corporate structure. This is particularly true of the larger producers. For example, in the corrugated container industry in 1969, 77.5 percent of all corrugated containers were produced by companies which combined mill operations and converting operations.[17] Since almost 90 percent of corrugated containers were produced by less than one-sixth of the companies manufacturing this product, it is obvious that it is the large producers who combine operations. In most cases of combined ownership, converting plants are located near board mills. Thus the residuals of the converting mills are very

[16] Initially, the General Services Administration included mill broke in their definition of recycled fiber as part of the President's policy to encourage recycling. This was intended to reduce the solid waste load, but it was obvious that it would have no effect on relieving the municipal solid waste burden, and the definition was rewritten on August 2, 1971.

[17] *Fibre Box Industry Statistics, 1969*, pp. 12, 14.

conveniently placed for recycling. They possess in a very high degree the quality of favorable location, one of the four qualities that determines whether or not residuals get recycled.

It is clear that the quality of mass is possessed by residuals in a converting operation, simply because the scale of a manufacturing operation is such that the generation of residuals is relatively large. The other two qualities of residuals required for efficient recycling are also inherently present to a high degree in converting operations. The scrap produced in a converting operation is homogeneous; it is all the same kind of material. It does not have to be winnowed out from a variety of other materials before it can be utilized. The scrap produced in a converting operation may also be said to be specifically contaminated, that is, the substances such as wax or glues or ink which have been added to the basic material in the processes of manufacture and conversion are known and are limited in extent.

The estimated 1 million tons of converting residuals not now recycled include the output of small widely dispersed producers such as printing job shops and small quantities of residuals contaminated with converting elements such as plastic inks or latex glues. The cost of transportation and processing may in these cases make recycling uneconomical.

The raw material characteristics of converting residuals have been sufficiently desirable so that more than 80 percent of this residual has been and is recycled year in and year out without any government regulation or incentives. The market as currently formed absorbs most of these residuals. In addition to the fact of a market demand for converting residuals, the cost of their disposal, if they are not recycled, is internalized. A plant generating the residuals must pay for hauling them away and incinerating or landfilling them if it does not sell or give them away for recycling. The desirable raw material qualities plus the internalization of disposal costs provide the incentive to recycle converting residuals. Most of them never reach the solid waste stream because of the circumstances of their generation.

Distribution Residuals

The third stage of the production–consumption sequence at which paper residuals are generated is distribution. This stage includes the obsolete and unsold inventories of paper and paper products. Probably the most significant component of it is overissue newspapers, which amounted to about 5

58

percent of production or about 425,000 tons in 1969. Some magazines are published with the anticipation that as many as 70 percent of the copies will remain unsold, but a national figure for this category of distribution residual is not available. Like converting residuals, overissue news has highly desirable characteristics as a raw material and generally is in demand in the paper stock market.

Consumption Residuals

Consumption residuals are generated at the fourth stage of the paper production–consumption sequence. These are the paper products themselves which, having fulfilled the function for which they were manufactured, are discarded, either to be stored in the environment until they degrade into compounds usable in nature, or to be recycled. It is estimated that in 1969 about 44 million tons of paper and paperboard products became solid residuals at the consumption stage. About 7 million tons or 14 percent of consumption residuals were recycled. About 37 million tons, or 87 percent, were discharged to the environment,[18] some with processing, such as incineration or grinding, as shown in figure 2, but most directly to landfill.

Consumption residuals are made up of two major categories, residential and commerical. The word *commercial* is used here in a very broad sense to include activities outside the home such as institutions, government, manufacturing, and wholesale and retail establishments. Of the 7 million tons of consumption residuals recycled in 1969, 1.6 million tons are estimated to have been residential consumption residuals—mostly newspapers—and 5.4 million tons are estimated to have been commercial consumption residuals—mostly corrugated shipping cases and mixed waste paper.

As the foregoing indicates, only if increased use of consumption residuals is achieved will recycling provide major assistance in solving the solid waste problem. In terms of desirable raw material characteristics, commercial consumption residuals present a more promising picture than residential consumption residuals, though increased use of the latter is possible without extensive modifications of present systems. For example,

[18] These estimates are based on data presented by Arsen Darnay and William E. Franklin, *Economic Study of Salvage Markets for Commodities Entering the Solid Waste Stream* (Washington, D.C.: U.S. Environmental Protection Agency, 1972), figure 13, p. 45–49.

separation and segregated storage of newspapers by householders, which is responsible for the major fraction of newspapers recycled today, can be increased. Other fractions of residential residuals—folding boxboard, brown kraft bags, corrugated boxes, printing and writing papers—are so lacking in mass and homogeneity that the potential for the recovery of these residential paper residuals appears to be lower than for the commercial consumption residuals.

Commercial Consumption Residuals

There appears to be a significant potential for increased use of paper residuals from commercial activities, including government offices. These activities are inherently on a larger scale than residential or domestic activities; consequently they generate reasonably homogeneous paper residuals in some quantity at a single location. It has been estimated that about 52 percent of all paper and paperboard consumption residuals are generated in residences and about 48 percent are generated in commercial activities.[19] On the basis of this estimate, commerce and industry, including government, are discarding between 15 and 20 million tons of paper residuals every year. This tonnage is a logical target for increased recycling because the scale and economics of the activities, and the centralized management of the activities involved, provide the framework for the generation and efficient handling of paper residuals with the product characteristics desirable for recycling.

Recycling is nothing new to prudent business managers. Converting residuals have been recycled for decades. Currently many large consumers of paper are reexamining their disposal practices. Wells Fargo Bank in San Francisco recovers for recycling every month on the average, 12 tons of tab cards, 6 tons of white ledger sorted from mixed waste, and 17 to 20 tons of mixed paper from the record storage department.[20] Throughout the economy, business managers in cooperation with paper stock dealers have been providing the 4 or 5 million tons of paper residuals recovered from commercial generating sources year in and year out. It is good business so long as there is demand for the material.

Table 4 recapitulates the generation of paper residuals at the various stages of production—consumption, the amount recycled, and the amount discarded in 1969.

[19] *Ibid.*
[20] Letter from Carroll H. George, Vice President, Wells Fargo Bank, to the author.

TABLE 4
Approximate Quantities of Paper Residuals Generated,
Recycled, and Discarded by Stage of Generation, 1969
(millions of tons)

Stage	Residuals generated	Residuals recycled[a]	Residuals discarded
Converting	5.5	4.5	1.0
Distribution	1.5	0.5	1.0
Consumption (residential)[b]	22.6	1.6	21.0
Consumption (commercial)	21.2	5.4	15.8
Total	50.8	12.0	38.8

[a]The total amount of paper recycled in 1969 was calculated by the Institute of Paper Chemistry on the basis of a survey of 800 paper mills. (W. S. McClenahan, "Consumption of Paper Stock by United States Mills in 1969 and 1970." Paper presented at the 1971 Secondary Fiber-Multi-Ply Board Joint Conference, Technical Association of the Pulp and Paper Industry.)

The allocation of recycled residuals to generation sources is supplied by Resources for the Future, Inc. The allocation of paper consumption to diverted and permanent use, and to residential and commercial consumption is adapted from calculations for 1967 paper consumption performed by the Midwest Research Institute. (Arsen Darnay and William E. Franklin, *Economic Study of Salvage Markets for Commodities Entering the Solid Waste Stream* (U.S. Environmental Protection Agency, 1972), figure 13, pp. 45–49.

[b]Total paper and paperboard consumption in 1969 was 58.9 million tons. The tabulation estimates 8 million tons went into nonpaper or permanent uses such as construction, permanent inventory, such as archives and business records, or was disposed of as liquid residuals.

Newsprint and Newspapers: The Residuals

The generation of newsprint residuals at the converting and distribution and consumption stages of the production–consumption sequence was depicted schematically in figure 2. Table 5 shows quantities of residuals generated at these stages in 1969, and their subsequent disposition.

For the 525 newspapers reporting to the American Newspaper Publishers Association (ANPA), converting residuals amounted to about 2.75 percent of tonnage run on the presses. The 525 papers accounted for 7.3 million tons in 1969. The quantity of converting residuals for the balance of newsprint uses—weeklies, commercial printers, shopping news, and comic magazines—is estimated at 10 percent.

Converting residuals reported in ANPA percentages include the wrappers that protect the jumbo rolls of newsprint in shipment, handling and transit

TABLE 5
Estimated Disposition of Newsprint and Newspaper Residuals in
1969 (thousands of tons)

Stage of flow	Residuals	Recycled	Addition to inventory	Final disposition
Conversion	442[a]	354[b]	–	88
Distribution	487[c]	390[d]	–	97
Consumption	8,812	1,418	717	6,676
Totals	9,741[e]	2,162[f]	717[g]	6,861

[a]Based on an average 2.75 percent converting loss for the 525 reporting papers of the American Newspaper Publishers Association plus a 10 percent converting loss for the balance.
[b]Estimate of 80 percent recovery is arbitrarily based on figures published by the printing industry.
[c]Based on a 5 percent return rate for all newspapers.
[d]Estimated of 80 percent of all distribution residuals.
[e]Total newsprint consumption of 9.741 million tons for 1969 is that reported by ANPA. The U.S. Bureau of Census reported 9.915 million tons consumed in this period.
[f]This quantity of newsprint recycled in 1969 is based on a thorough and knowledgeable survey of all paper recycling in 1969–70 by the Institute of Paper Chemistry in Appleton, Wisconsin. Total paper recycled is reported as 11.969 million tons. Prior to the publication of this survey, the American Paper Institute's *Statistics of Paper* used the Bureau of Census figure of 10.939 million tons. The discrepancy continues in 1970 with API showing 12.621 against Census's 10.530 and in 1971, API showing 13.156 against Census's 10.265.
[g]Addition to inventory includes quantities of newsprint used for gypsum wallboard facings and other construction applications.

waste, white press (newsprint spoiled on the roll but not printed), core waste (the paper left in the jumbo roll when it is removed from the presses and a new roll put on), and printed waste.

The balance of newsprint residuals, 8.8 million tons, is shown as consumption residuals in table 5. A newspaper is one of the more ephemeral products of our society. Candy wrappers have a comparably short life after final distribution has taken place, but the time span between manufacture and final distribution may be several months. On the other hand, newspapers are manufactured and distributed several times a day, and the time from manufacture to final purchase and use may be less than an hour, as when a paper just delivered to a street vendor is sold to a commuter who discards it on the bus at the end of his ride.

For purposes of analyzing solid residuals management, consumption residuals are classified according to generating locations as commercial residuals and household residuals. It has been estimated that 92 percent of all persons who read newspapers daily read them at home, 7 percent at work, and 3 percent elsewhere. Of the people who read two or more papers a day, 94 percent read them at home, 13 percent at place of work, and 6 percent elsewhere.[21] These figures show that newspapers are primarily a residential rather than a commercial residual.

The scale of generation of newspapers as a residential residual may be determined by examining the density of households—whether in multiple or detached dwellings—and such characteristics of the householders as income level and education. A profile of newspaper readership developed from an income dynamics study of 148,288 households by the Survey Research Center of the University of Michigan in 1969[22] indicates that about 88 percent of persons living in households with family incomes of $15,000 or more read a newspaper every day while less than 22 percent of persons in households with family incomes of $5,000 or less read a paper every day. Where configuration of dwelling units is the variable, the highest readership, 78.4 percent, occurs in apartment houses of more than three stories. Of the persons living in detached single-family houses, 73.6 percent read newspapers every day. The lowest percentage of daily readership, 57.5, occurs in families living in trailers. When education of the family head is considered, 84.3 percent of people in the sample with a bachelor's degree read a paper daily compared to 78 percent whose education did not continue beyond high school. These data may be helpful in assessing where, in a city, concentrations of newspaper residuals are apt to be found.

Disposition of Newspaper Residuals

Table 5 shows about 6.9 million tons of newspaper residuals discarded in 1969. This quantity is reflected in several studies which show newspaper residuals as 10 to 14 percent of collected municipal solid residuals.[23] A moment's reflection will indicate that these percentage figures imply a

[21] W. R. Simmons Co., "Selected Markets and the Media Reaching Them, 1970." *Newspaper Audience Report,* table 22 (New York).

[22] Survey Research Center, University of Michigan, special computer run of data for the author, March 1971.

[23] A figure of 14 percent is cited as an average in *Proposals for a Refuse Disposal System in Oakland County Michigan* (U.S. Dept. of Health, Education and Welfare, 1970), p. 9. A slower figure of 9.4 percent is cited in E. R. Kaiser, "Chemical Analyses of Refuse Components," *Proceedings 1966 National Incinerator Conference* (New York: American Society of Mechanical Engineers, 1966), pp. 84–88.

range of 49 to 69 million tons for total municipal solid residuals, a figure well below the 150–200 million tons variously cited as being total municipal solid residuals. It is probably more accurate to say that newspapers account for between 10 and 14 percent of the residuals collected from urban residences, not including commercial, industrial, and street cleaning residuals, which comprise a part of total municipal residuals.

Newspapers have some good qualities and some poor qualities as disposable residuals. When collected flat and folded, they have a relatively high density. They are, however, somewhat resistant to disposal. Any housewife whose husband borrows the sewing shears to clip his newspaper knows of the unsuspected abrasive quality of paper. It is a component of MSR that has a more dulling effect on the knives of shredders, hoggers, and grinders than most other residuals. Furthermore, newspaper does not incinerate readily. Because of its density in a flat, folded state, it may pass through an incinerator with an 1,800° burning temperature and still leave some readable pieces in the quenched residue. Newspaper that is directly landfilled may still be readable 20 years from now if it has not encountered the right combination of moisture, bacteria, chemicals, and oxygen.

The newspaper residuals that were not disposed of in 1969 but were recycled instead amounted to 2.2 million tons, as shown in table 5. Of this total, 1.5 million tons were consumption residuals, while the balance of 0.7 million tons was generated at the converting and distribution stages.

The end products for which the 2.2 million tons of newsprint and newspaper residuals supplied the raw material are shown in table 6. Sixty-seven percent of recycled newspaper residuals was used in the manufacture of paperboard, primarily boxboard; 19 percent in the manufacture of paper, mostly newspaper; and 14 percent was used for other paper and paperboard, primarily construction paper and paperboard.

Generation and Disposition of Corrugated Container Board Residuals

The manufacture of a corrugated shipping container involves six different operations which may be arranged in a variety of physical structures. There are two pulping operations, one for linerboard and one for the corrugating medium; two paperboard manufacturing operations, one each for the linerboard and corrugating medium; then there is the operation in which the corrugating medium is fluted and glued to the liners to make sheets of corrugated container board; and finally the scoring, slitting,

TABLE 6
End Uses of Recycled Newsprint and Newspapers in
1969 (thousands of tons)

Paper			
Printing	368		
Packaging and industrial converting	3		
Tissue	32	Total paper	403
Paper board			
Boxboard	1,050		
Container board	3		
Packaging and industrial converting	319		
Classification not established	72	Total paperboard	1,444
Other Paper and Board			
Construction	198		
Molded pulp and miscellaneous	117	Total other	315
			2,162

Source: W. S. McClenahan, "Consumption of Paper Stock by United States Mills in 1969 and 1970." Paper presented at the 1971 Secondary Fiber-Multi-Ply Board Joint Conference, Technical Association of the Pulp and Paper Industry.

slotting, gluing or stitching and printing required to make the completed shipping container. The last two operations comprise what were defined earlier as converting operations.

In 1969 over 15 million tons, more than one quarter of all paper and paperboard produced, consisted of linerboard and corrugating medium for use in making shipping containers. These products were shipped from their separate manufacturing operations to box plants where they were made into about 13.4 million tons of corrugated container board, generating 1.7 million tons of converting residuals in the process. Used corrugated containers are estimated to have been 13.4 million tons in 1969, the same quantity as produced. This estimate assumes that quantities in inventory at all stages of the flow were equal at the beginning and end of 1969.

The sites where corrugated containers generally become consumption residuals were noted above. Large-scale sites tend to be those where goods are unpacked to be sold in smaller lots or to be used in assembly plants. Smaller scale sites include such activities as office buildings and restaurants where goods are delivered in wholesale lots for consumption over a period of time.

Of the 13.4 million tons of corrugated container consumption residuals generated in 1969, 2.6 million tons were recycled and 10.8 million tons went to incineration and landfill. The end products into which these residuals were recycled are summarized in table 7. The most significant end uses shown in the table are boxboard and shipping container board, the latter being the same product that is the source of the corrugated container residual. The third largest use, packaging and industrial container board, includes such items as tube and drum stock and lining for gypsum wallboard.

Raw Material Qualities of Corrugated Container Residuals

The qualities of residuals that give them value as raw material have been examined previously for paper in general and for newsprint in particular. Concentration of mass, homogeneity of the material, and a low level of contamination are three of the qualities over which the generator may have some control.

The homogeneity of the corrugated container residuals is rather easily established by visual identification. It is not possible to ascertain that two pieces of steel are alike by simply looking at them because different combinations of alloys may present the same appearance while being quite different in substance. Color, shape, thickness, and the fluted middle of the sandwich all identify corrugated container board and make it readily distinguishable from other paper and paperboard residuals. Liners and corrugating medium may be specially treated or coated in the course of manufacture to make containers waterproof, and to make the inside liner nonabrasive for packing china or to keep contents such as uncured rubber from sticking to them. Coatings of wax, plastic, and waterproof starches are used for these purposes. Successful recycling of specially treated corrugated containers depends on separation according to the coating used. The different kinds of coatings can generally be determined visually or by knowledge of the type of product that has been packed in the container. The great majority of corrugated containers are not coated or specially treated, so the level of this quality in the DC/SMSA area is not critical to an increased level of recycling.

Homogeneity requires separate handling and storage of the residual at the site of generation. It may well be to the economic advantage of the residuals generator to provide separate handling and storage for corrugated

TABLE 7
End Use of Recycled Corrugated Container Board
Residuals in 1969 (thousands of tons)

Product	Quantity	
Paper		
Printing	8	
Packaging and industrial converting	8	
Special industrial	1	
Tissue	58	
Total paper		75
Paperboard		
Boxboard	1,336	
Shipping container board	1,665	
Packaging and industrial converting	699	
Classification not established	455	
Total paperboard		4,155
Other paper and board		
Construction	207	
Wet machine	9	
Molded pulp and miscellaneous	4	
Total other		220
Total all uses		4,450

Source: W. S. McClenahan, "Consumption of Paper Stock by United States Mills in 1969 and 1970." Paper presented at the 1971 Secondary Fiber-Multi-Ply Board Joint Conference, Technical Association of the Pulp and Paper Industry.

container board residuals. Because of the low density and high strength of corrugated container board and the bulkiness of corrugated containers, it is expensive to discard them. The average set-up corrugated container takes up so much space in residuals storage that frequently people collapse them prior to collection. Collapsing containers manually is an awkward task which involves unsticking the bottom panels that have been taped, glued, or stitched together, and then folding them to their original flat form. A conservative estimate of the labor cost of manually collapsing a ton of used corrugated containers is $60.

The cost of hauling corrugated residuals even if they are collapsed is relatively high because of their low density. The compaction ratio for corrugated container board residuals in a packer truck is less than one-half the ratio for MSR.[24] Furthermore, corrugated residuals, more than other residuals, have a tendency to spring back from a compacted state after compression has been removed; the effort of compaction does not remain effective unless the corrugated container board residuals are tied together under compression, as when they are baled.

These characteristics of corrugated tend to enhance its recyclability in a kind of perverse way. The general argument against recycling and against separation of residuals at source to make them more recyclable is that separate handling and collections are prohibitively expensive. The costs involved cannot be recovered in the value of the salvaged material. However, corrugated requires special processing as a discard residual simply because of its bulk. If the processing costs for discard approach the processing costs of saving for recycling, and they appear to, then the value of the salvaged material, however small, exceeds the value of the discarded material.

The extent and variety of contamination, a second residual quality, is determined partly in the manufacturing process, partly in use, and partly in handling following use. Printing ink added during the converting process may make recycling difficult, particularly if the ink has a plastic base. Closures such as staples, pressure-sensitive glues, and asphalt-laminated sealing tapes are all contaminants added to the corrugated container board in the course of use. These present varying degrees of difficulty to the recycling mill. The product contents of a corrugated container board may be another source of contamination, particularly if the material stains, or soaks into, or adheres to the liner. Finally, the handling of the corrugated container after it becomes a residual may determine whether it is too contaminated to recycle. Corrugated containers mixed with other solid residuals are not readily separable and cleanable for recycling. The varieties of contamination present in an aggregation of mixed solid residuals may be the most troublesome because of the need for what might be called "broad spectrum" cleaning rather than specific

---

[24] Conversation with Al Shayne, Shayne Bros. Trucking Co., Washington, D.C. The fluting of the center layer of corrugated container board encloses large volumes of air space that remain even after a shipping container has been collapsed or folded to a flat shape. Normal expectation is that a loose yard of 180 pounds of household residuals occupying 1 cubic yard can be compressed in a packer truck so that 540 pounds of residuals occupy a cubic yard. If only corrugated container residuals are loaded, 180 pounds can only be compacted to a volume resulting in a density of 270 pounds per cubic yard.

cleaning. Corrugated containers mixed with other residuals lose the quality of uniformity of contamination and as a result become more difficult to recycle.

Concentration of mass for corrugated container residuals is achieved partly as a matter of scale and partly as a matter of the same economics of handling that promote homogeneity. A great number of corrugated container residuals are generated in the course of a day's sales from a supermarket or a liquor store. Efficient handling of these residuals, whether in preparation for recycling or for discard, requires that their volume be substantially reduced either by hand labor or by baling. The cost of manual labor in most circumstances is prohibitive.

There are a great variety of balers available for this operation, ranging from a portable, hand-operated, string-tied bundle type that turns out 45-pound bales and costs about $400, to an electrically powered hydraulic unit that turns out 200-pound bales and costs about $3,000. Larger units are available for making bales from 500 to 2,000 pounds.

Corrugated container residuals in a set-up state have a density as low as 20 pounds per yard.[25] If they are loaded into a packer truck without further processing, the density can be increased to 100 pounds per cubic yard. The density achieved on the small balers noted above ranges from 135 pounds per cubic yard to 540 pounds per cubic yard.[26]

Relative Cost of Paper Residuals and Virgin Pulpwood

About 20 percent of the paper residuals generated are currently recycled—12 million tons out of 58.9 million tons consumed in 1969. This proportion compares with 35 percent toward the end of World War II and over 30 percent in the 1920s. To understand the present market for paper residuals, it is necessary to look at some of the factors affecting the relative costs of paper made from paper residuals and paper made from virgin wood. The most significant factors are raw material availability, the organization of the paper industry, transport, product specifications, and residuals management.

[25] This is readily established by calculating the number of containers with a capacity of about 1.5 cubic feet and a weight of 1.3125 pounds that will occupy a cubic yard.
[26] W. M. Russell, "Study of Baling Equipment and Systems Employed in Retail Stores," a report prepared for the Boxboard Research and Development Association, March 1970.

The rapid decline in the use of paper residuals to produce paper follow-ing World War II coincided with the development of technology for pulping highly resinous softwoods, and previously unpulpable hardwoods, and the first extensive pulping of wood products residuals. An improve-ment in the sulfate process expanded the species base of pulping in both the Southeast and the Northwest. Later came the introduction of the neutral sulfite semichemical pulping process, which provides a sub-stantially higher yield than full chemical pulping processes. The neutral sulfite semichemical pulping process also yields a cooking liquor residual that is usable in kraft sulfate pulping. The pulping of sawdust and other wood product residues was made possible by the continuous digesting process of pulping and as a result of management initiatives to reduce raw material costs. By 1970 about 67 percent of the raw material input to paper production were forest products residuals and 6 percent were from lumbering, bringing the total to 73 percent.[27]

The impact of wood pulp made available by technological innovation shows up vividly in the statistics of container board production, which was and is the largest product consumer of sulfate and semichemical pulps. In 1943–44, paper residuals provided 40 percent of the raw material input used in the manufacture of 4.2 million tons of corrugated container board.[28] In 1970, paper residuals made up less than 10 percent of the input in the manufacture of 13.7 million tons of corrugated board.[29] At the same time that these technological innovations were decreasing the cost of pulp made from round wood, increasing labor costs for the collection and sorting of paper residuals was making waste paper more expensive as a raw material. Also, there was a marked increase in additives in paper products, making the reclamation of the cellulose fibers more difficult and hence more expensive.[30] However, the projected doubling in demand for paper and paper products to over 100 million tons in 1985, the competing demands for forests as recreation and wilderness areas, and

[27] Ronald J. Slinn, "Availability of Residuals in U.S. Pulp and Paper Industry," chart 8. A paper presented at the 77th National Meeting of the American Institute of Chemical Engineers, Pittsburgh, Pa., 1974.

[28] Dwight Hair, *Use of Regression Equations,* table 13, p. 156.
Paper and Board, U.S. Department of Agriculture, Forest Service Resource Report No. 18, December 1967, table 13, p. 156.

[29] J. Rodney Edwards, "What's Ahead in Waste Paper Utilization," speech, American Paper Institute, March 15, 1971.

[30] As noted below, the relative costs did not include all of the relevant residuals management costs for either raw material.

the demand for wood fibers for construction materials are combining to modify the relative costs of round wood and paper residuals as raw materials for paper production. These factors point to a rising cost for round wood.

The organization of the paper industry appears to be a factor favoring pulpwood over paper residuals as a raw material. Frequently the raw material source, the forest, is owned by the pulp manufacturer. The pulp mill may be integrated with the paper mill, and the paper mill may be combined with a converter such as a box plant. This arrangement, complete vertical integration from forest through final product, has potential economies for management overhead, capital investment, taxes, and transportation. The extent of integration of pulp mills with paper mills in the United States is indicated by the fact that of the 44 million tons of wood pulp consumed by U.S. paper and board mills in 1969, 36 million tons were intracompany transfers and only 8 million tons were purchased from other domestic and foreign companies.[31]

There are some instances of paper mills owning waste paper dealerships, but this structure for the recycling of paper residuals is not parallel with the integration of virgin paper mills. In the first place it is on a much smaller scale. That is to say, the proportion of paper stock dealers owned by paper mills that use paper stock as a raw material is much smaller than the proportion of pulp mills owned by paper mills using virgin wood pulp as a raw material. In the second place, the integrated recycled paper mill does not have control over its raw material source (paper residuals) to the extent that the virgin paper mill does over its raw material source, the forest. The trees in the virgin paper mill's forest lands are a known quantity, in existence well ahead of their planned use. The only real hazards to production are fire and flood. The availability of paper stock to the recycled paper mill, on the other hand, depends on the behavior of the sources that generate the paper residuals and the efficiency of the operation that collects them.

The effect of transport costs on the relative costs of paper residuals paper[32] and virgin wood pulp paper is a function of both the freight rates charged by railroads and truckers and the distances that raw material inputs and product outputs have to be shipped.

[31] U.S. Department of Commerce, *Quarterly Industry Report,* "Pulp, Paper, and Board," October 1970.
[32] Paper residuals paper is an awkward phrase coined to designate paper manufactured from the raw material of paper residuals.

71

There has been considerable discussion over whether rates charged for transporting paper stock are unfair and discriminatory, giving pulpwood an unfair price advantage as a raw material. This is a complex question that has not been answered, and it is not proposed to try to answer it here. There are differences in the location patterns of virgin wood pulp paper manufacturers and paper residuals paper manufacturers which indicate relative cost advantages arising from distances that materials have to be shipped. Some conclusions can be drawn here from these patterns.

About four-fifths of U.S. paper manufacturing capacity is designed for virgin wood pulp paper. This capacity cannot utilize paper residuals as raw material because it is not equipped with the required cleaning and deinking equipment. By and large this capacity has been located next to forests and away from cities; thus it is adjacent to its raw material supply and does not have to pay costs for shipping raw material great distances. The availability of raw material is not a constraint on the size of the plant capacity. Whether the plant is designed to produce 300 tons a day, 600 tons a day, or 1,000 tons a day is a question answered by considerations other than the availability of raw material. Being located away from the centers of population, the virgin raw materials plant does incur relatively greater costs for the distance its finished goods must travel to market.

Secondary fiber mills[33] designed to utilize paper residuals as raw material are located in or next to the cities which supply their raw material. They have the advantage of being close not only to their sources of raw materials but also to the markets for their finished products. But there are severe limits here on manufacturing capacity that arise from the daily or weekly availability of adequate supplies of raw material. There are minimum daily and weekly production capacities below which paper mills, as capital-intensive activities, cannot economically operate. If the adjacent population center generates enough homogeneous paper residuals to meet the minimum daily raw material inputs of a secondary fiber mill, then the mill realizes fully the advantage of being next to the population center. But, to the extent that the mill must go farther afield for raw material supplies, to this extent is the location advantage diluted, and there is a point, depending on the value of the raw material input, beyond which it is uneconomical to ship paper residuals as a raw material.

[33] The term *secondary fiber mill* is synonymous with the term *recycling mill*. Secondary fiber is cellulose fiber that has been made into paper one or more times.

Effect of Residuals Management on Relative Costs

The imposition of quality standards on the gaseous and liquid residuals discharged to the environment by paper mills increases the processing costs of both virgin pulp paper and paper residuals paper. Steps must be taken in processing to prevent the discharge of gases and particulates up the stack and to dispose of them by incineration or landfill in an acceptable manner. Likewise, processing must be added to remove from sewage discharges the chemicals and solids prohibited by the quality standards. This requirement for residuals management internalizes the cost of residuals generation in the paper industry. "The primary residuals generated in (virgin) pulp and paper production are gaseous residuals from chemical pulping (digestion)—sulfur dioxide, and to a much lesser extent particulates, from sulfite processes; organic sulfides (odorous compounds), and particulates from the kraft process; dissolved inorganic and organic solids from bleaching and digestion operations; bark and dirt from wood preparations; and fiber primarily from machine white water."[34] In the past these residuals have been discharged to the water courses, the atmosphere, and the land with little regard for environmental effect. The imposition of standards of environmental quality has required that manufacturers utilize more extensive techniques of residuals management to maintain environmental quality. These techniques include such operations as removal of particulates going up the stack, use of low-sulfur fuel to reduce sulfur dioxide discharge, installation of sedimentation basins and aerated lagoons for the settling out of liquid residuals, and development of incineration and sanitary landfills to provide for solid residuals. Residuals management may also result in changes in the basic manufacturing process to reduce a particularly bothersome residual. The addition of these elements to the manufacturing processes of virgin pulp paper tends to increase the cost of this paper.

On the other hand, the manufacture of paper residuals paper is not free of the cost of residuals management. The costs may be relatively greater or less than those of virgin pulp paper, depending on the raw material inputs and the required specifications of the final product, particularly the level

[34] "Residuals Generation in the Pulp and Paper Industry," Blair T. Bower, George O. G. Löf, and W. M. Hearon, *Natural Resources Journal,* Vol. 11, No. 4, 605–623, 1971.

of brightness to which the final product must be bleached. Paper residuals may have coatings or fillers which have to be removed in recycling in addition to ink. The repulping of cellulosic fiber reduces fiber size, increasing the amount of fibers lost in the papermaking process. Brightness specifications necessitate the use of bleaching chemicals. All of these materials become liquid residuals in the recycling process and have to be removed from mill effluent in accordance with environmental quality standards.

While recycling mills may be faced with lesser gaseous effluent problems than virgin pulp paper mills, the quantity of liquid and solid residuals to be disposed of per ton of input may be considerably greater than for virgin paper mills. One study has shown the cost of residuals removal for certain grades of recycled paper to be several times greater than the cost for virgin pulp paper, depending on the level of production.[35]

There are costs, currently not assessed to any operation, associated with the discharge into the environment of the residuals that remain even after process changes are made to meet environmental quality standards. These costs are difficult to quantify. They encompass the damages, if any, resulting from the remaining SO_2, particulates, and reduced sulfur compounds discharged to the atmosphere; the damages, if any, of suspended and dissolved solids discharged to the water courses; and the externalities associated with landfill operations.

It should be noted that this overview of some of the factors bearing on the relative costs of pulpwood and waste paper as raw material for the manufacture of paper and paperboard products does not include discussion of the competition between products of different materials capable of fulfilling the same function, and therefore is somewhat incomplete. Over the years paperboard containers have largely replaced wood boxes, wire-bound boxes, and barrels as containers for meat and produce. Currently, plastic is competing with boxboard in the manufacture of set-up boxes and in other paper applications. For the purposes of this discussion we must assume that wood fiber will not be supplanted by substitute materials as the prime raw material for paper. Thus the discussion of relative costs can be limited to virgin wood fiber versus recycled wood fiber without consideration of possible substitutes.

The declining percentage of recycled paper has been, in large part, a function of the relative costs of the two raw materials—paper residuals and

[35] Bower, Löf, and Hearon, manuscript in process.

74

pulpwood. The latter has had an advantage over waste paper in the areas of raw material availability, the organization of manufacturing facilities, and in some cases transportation. In the fourth area, residuals management, increasingly stringent environmental controls are requiring more expenditures. It appears that in many cases these costs are larger for pulpwood than for paper residuals as a raw material.

Demand for Recycled Paper

The behavior patterns expressed in the demand for paper affect the extent to which paper residuals are used. There are three different markets in which demand is expressed for recycled or virgin cellulosic fiber. The first is the raw material market where there is a choice of paper residuals or round wood, or wood product residues, or a combination of the three, for manufacturing paper. The short-run demand in this market, as indicated above, tends to be a function of location and production facilities. The second market is for paper as a producer good, that is, it is used to package another product, or it is used to convey information in the form of a book, a newspaper, a brochure, or a greeting card. The purchaser is commerce or industry. Product preference here is not a direct expression of the private consumer, but the private consumer has a substantial influence on the preferences that are expressed. A prime example is the current movement of governments and corporations to purchase paper products made from recycled fiber as a means of soliciting the goodwill of citizens, as well as a means of reducing solid residuals. The third market is the private consumer market. The proportion of paper production distributed on this market is relatively small—such products as toilet and facial tissue, napkins, towels, so-called disposable products, and stationery. The demand of the private consumer for these goods is expressed directly. In the past, the opportunity to express a preference between recycled and virgin pulp material was limited by lack of product material identification. Now many suppliers seek to exploit public concern for the environment by labeling their paper products as recycled.

In all three of these sectors of the paper products market, demand appears to be a function of socio-psychological factors as well as of price and quality. One example of this is found in the language of the paper trade. Paper made from recycled fibers has in the past been referred to as "bogus," defined by the dictionary as a synonym for counterfeit. *The*

Dictionary of Paper[36] defines bogus as "a descriptive term applied to papers and paperboards manufactured principally from old papers or inferior or low grade stock in imitation of grades using a higher quality of raw material." The prejudices of a culture using this vocabulary do not allow for the possibility that waste paper could be a raw material of quality equal to virgin pulpwood under any circumstances. The necessary consequence of this bias is to assign a low social value to recycling. In the past, this set of attitudes has resulted in customary, explicit specification of virgin pulp paper, a specification not necessarily related to the intended use of the paper or to its performance characteristics.

Function of the Paper Stock Dealer

The secondary materials dealer has been a vital cog in papermaking for over a hundred years, from the time when all paper in the United States was made from cotton rags, to the present. Some of the larger paper companies which own both mills and converting operations may recycle converting residuals without going through a dealer. But by far the greatest percentage of waste paper recycled is a purchased commodity.[37]

Paper residuals prepared for sale to a paper mill are known as paper stock. The entrepreneurs who perform this operation are known as paper stock dealers. The paper stock dealers and users of paper stock have an association, The Paper Stock Institute of America, which is a division of the National Association of Secondary Materials Institute. Paper stock dealers divide into three groups—brokers, wholesalers, and retailers. In many instances wholesalers also double as brokers. As in any other commodity trading, a paper stock broker is knowledgeable in what buyers are buying and what sellers are selling. He brings buyers and sellers together and for this service receives one or two dollars per ton. He may augment this activity by hauling the paper stock himself as a private carrier.

The words *wholesaler* and *retailer* differentiate amounts purchased rather than amounts sold. A wholesale paper stock dealer purchases all the

---

[36] 3rd ed., p. 75.

[37] For the most part, wood pulp is not a market commodity as waste paper is because most wood pulp is manufactured in mills owned by paper producers. Wood pulp sold outside the integrated operations is designated market pulp. The proportion of total purchased waste paper "going through" paper stock dealers is not known.

paper residuals from a large printing facility, for example, while a retail paper stock dealer purchases relatively small amounts of paper from householders and others and sometimes sells his output to the wholesaler.

The aggregation of paper residuals for processing into paper stock occurs in a variety of ways. Converting residuals of large industrial generators are generally sold under a contract whereby the paper stock dealer agrees to buy all the paper scrap of a plant of a specific description and packed in a specific way for a definite price over a stated period of time, generally a year. The price includes consideration of who transports the paper residuals to the dealer's dock, the seller or the dealer.

Commercial consumption residuals may be sold in the same way by companies that have developed programs for providing segregated storage of their paper residuals. The Wells Fargo Bank in San Francisco, mentioned earlier, is an example. Many supermarket distribution centers follow this practice.

In some cases the commercial collector of mixed solid residuals is more sophisticated in matters of solid waste management than the residuals generator. When he is aware that he has a load that has a considerable amount of clean paper residuals, such as corrugated containers, or large quantities of computer printout paper or tabulating cards, the collector delivers his load to a paper stock dealer who buys the good paper residuals the collector picked up as mixed solid residuals destined for landfill.

In addition to collecting and receiving waste paper, the paper stock dealer sorts and grades it. The tremendous variety of paper products and the number of pulp combinations associated therewith create a large amount of ambiguity and confusion in the paper stock market. To reduce this confusion to the lowest possible minimum and enable transactions to take place without prior inspection of material, the Paper Stock Institute of America publishes yearly a circular of *Paper Stock Standards and Practices*.[38] The edition for 1972 defines forty-six different grades of paper stock which a dealer may pack, and sets forth the specifications governing each grade. The great majority of paper stock grades, almost forty, is made up of converting residuals. Only five or six grades cover paper that has gone through the consumption stage of the production—consumption sequence. Among the converting residuals grades are cor-

---

[38] Paper Stock Institute of America, a Commodity Division of National Association of Secondary Material Industries, Inc. *Paper Stock Standards and Practices,* Circular PS-69, New York, 1969.

rugated cuttings, white news blanks, No. 1 soft white shavings, hard white shavings, and hard white envelope cuttings. The consumption residual grades include No. 1 news, corrugated containers, sorted white ledger, and manila tab cards.

The grades serve as commodity descriptions on which regional prices are quoted both for the paper residuals sold to the paper stock dealers and for the sales of paper stock to paper mills. Some illustrative prices for paper stock quoted on the East Coast f.o.b. dealer's plant in September of 1971 were $95 to $100 for hard white envelope cuttings, $42.50 to $45 for sorted white ledger, and $15 to $17 for No. 1 news.[39]

The grade descriptions in *Paper Stock Standards and Practices,* which is issued anew every year, in effect act as the specifications for the different kinds of paper stock. For example, the specification for grade 2, No. 1 mixed paper, states "Consists of a mixture of various qualities of paper packed in bales weighing not less than 500 pounds and containing less than 25 percent of soft stocks such as News." "Prohibitive materials" are limited to 1 percent and "Total outthrows" are limited to 5 percent. Prohibitive materials are defined as materials that may be damaging to equipment; hot melt glues are an example. They are also materials that make the pack unusable in the grade specified. Outthrows are defined as papers in a form that is unsuitable for consumption in the grade specified. Carbon paper is an example of an outthrow.

The five grades of paper stock covering newsprint and newspaper residuals are listed in table 8. Grades 6 and 7 can be seen from the definitions to be consumption residuals. Grade 8 comprises distribution residuals, and grades 24 and 25 are converting residuals.

Table 9 is a listing of the paper stock grades and East Coast prices for the five different grades of corrugated container residuals that paper stock dealers trade in. It shows the 1.7 million tons of converting residuals and the 2.6 million tons of consumption residuals used in 1969. Grade No. 10, corrugated containers, is the lowest grade. This is a consumption residual. The liners (outside layers) may be of either jute (recycled fiber) or kraft (at least 85 percent virgin fiber). Grade 10 is quoted $2.50 below grade 11, new corrugated cuttings, a converting residual. According to the grade description, the cuttings may be jute or kraft, but may not contain any printed material.

---

[39] *Official Board Markets,* Magazines for Industry, Chicago, September 4, 1971, p. 3.

TABLE 8
Paper Stock Grades Describing Newsprint and
Newspaper Residuals

Grade No.	Description
(6) No. 1 news	Consists of newspapers packed in bales of not less than 54 inches in length, containing less than 5 percent of other papers. Prohibitive materials may not exceed 1/2 of 1 percent. Total outthrows may not exceed 2 percent.
(7) Super news	Consists of sorted fresh newspapers, not sunburned, packed in bales of not less than 60 inches in length, free from papers other than news and containing not more than the normal percentage of rotogravure and colored sections. No prohibitive materials permitted. Total outthrows may not exceed 2 percent.
(8) Overissue news	Consists of unused overrun regular newspapers printed on newsprint, baled or securely tied in bundles, and containing not more than the normal percentage of rotogravure and colored sections. No prohibitive materials permitted. No total outthrows permitted.
(24) White news blanks	Consists of baled unprinted cuttings and sheets of white newsprint paper or other papers of white ground wood quality. No prohibitive materials permitted. Total outthrows may not exceed 1 percent.
(25) Super white news blanks	Consists of baled unprinted cuttings or sheets of white newsprint of uniform brightness and quality. No prohibitive materials permitted. Total outthrows may not exceed 1/2 of 1 percent.

Source: Paper Stock Standards and Practices Circular PS-69 (New York: National Association of Secondary Materials Industries, Inc., 1969).

Grade 12, new double-lined corrugated cuttings, mixed medium, at
$27.50 is $5 a ton higher than the corrugated cuttings (grade 11) because
the liners are specified as all kraft. Jute (recycled) liners are not acceptable
in this grade. The corrugating medium, however, may be of recycled fiber.

Grade 13, new kraft-lined corrugated cuttings, with a 9-point semichemical or kraft corrugating medium at $32.50, is $5 a ton more than grade 12,

TABLE 9
Quantity and Price of Grades of Corrugated
Paper Stock Recycled in 1969

Grade No.[a]	Description	Quantity[b] (000 tons)	Price per ton[c]
(9)	Solid fiber containers Consists of solid fiber containers having liners of either jute or kraft packed in bales.	43	Not quoted
(10)	Corrugated containers Consists of corrugated containers having liners of either jute or kraft, packed in bales of not less than 54 inches in length.	2,587	20.00
(11)	New corrugated cuttings Consists of baled corrugated cuttings having two or more liners of either jute or kraft. Butt rolls, slabs of medium, and pointed containers are not acceptable in this grade.	405	22.50
(12)	New double kraft lined corrugated cuttings Consists of baled corrugated cuttings have all liners of kraft. This grade shall be free from nonsoluble adhesives, butt rolls, slabs of medium, and printed containers.	1,167	27.50

Continued

80

TABLE 9
(Continued)

Grade No.[a]	Description	Quantity[b] (000 tons)	Price per ton[c]
(13)	New brown kraft corrugated cuttings Consists of baled corrugated cuttings having all liners of brown kraft. The corrugated medium must be either semi-chemical or kraft. This grade shall be free from nonsoluble adhesives, butt rolls, and pointed containers.	76	32.50
	Total	4,278	

[a]Grade numbers and descriptions come from *Paper Stock Standards and Practices, 1969.*
[b]Quantities were given to the author in a personal communication from W. S. McClenahan of the Institute for Paper Chemistry in Appleton, Wisconsin. These figures represent a later analysis of data than the figures in the table; thus the difference of 122,000 tons in total quantity may be ascribable to a modification of the 455,000 tons whose classification was not established in table 7.
[c]Prices are from *Official Board Markets,* June 5, 1971.

primarily because the corrugating medium will be of virgin pulp of a specific kind rather than recycled fiber. This grade must also be free of nonsoluble adhesives and printing.

The separation of the converting and consumption residuals of corrugated containers into four grades distinguished as (a) used, (b) new, (c) all kraft liners, and (d) virgin pulp corrugating medium is illustrative of the complexity that enters into the grading operation of the paper stock dealer. It also is a striking example of the importance of maintaining identification on the basis of separation and knowledge of residuals generating source.

The sorting function of the paper stock dealer has diminished markedly over the years because the cost of labor has increased relative to the value of paper stock. Some people attribute the decline of the paper recycling ratio, in good part, to this decline. For the most part, mixed papers of a diverse furnish and production process, including hard-to-handle contaminants such as plastic covers and carbon paper, will not be accepted by a

81

paper stock dealer today if sorting is necessary to achieve salability. Some mixed paper free of certain classes of contaminants can be marketed to construction materials firms without sorting. Mixed loads of paper residuals that come into a paper stock dealer with large amounts of homogeneous paper may be worth separating.

Some cleaning operations may be carried out by paper stock dealers, particularly on converting and distribution residuals. For example, overstocks of paperbound pamphlets may be placed under a mechanical guillotine-type cutter to trim off binding made with hot-melt glue, and the covers may be separated from the text and packed with a different grade of paper stock. Whether or not an operation such as this is economical depends on the value of the resulting paper stock and the cost of the operation.

The final operation, sometimes the only operation, carried on by the paper stock dealer is packing. This operation is extremely important because of the high ratio of handling cost per unit of material to value per unit of material. The preparation of dense, uniform, strong, and properly dimensioned bales is essential today to the marketing of paper stock. The maximum possible density of paper stock is sufficiently low so that the limiting factor in both truck and rail car loading is the cubic capacity of the vehicle rather than maximum allowable weight per axle. Bales of maximum density, designed to stow in the greatest number in truck and freight car, are necessary. Sturdy bales weighing about a ton and of uniform shape are desirable so they can be handled and stored by fork-lift trucks. Generally, bales of the desired density cannot be achieved without reducing the size of the paper fragments being packed; consequently, many dealers, prior to baling the paper, put it through a hogger that tears it into small pieces. The conveyor, hogger, and baler required to turn out this type of pack costs on the order of $150,000.

Aggregating, cleaning, sorting, grading, and packing paper stock are the operations carried out by the waste paper dealer. These operations are analogous to the processing of round wood in the forest into pulpwood. Paper stock does not require a chemical cooking process to separate out unwanted wood tissues, but it must go through a cleaning cycle in the paper mill in which all of the contaminants are removed. The process may involve the dispersion of asphalt or the deinking of printed paper or simply the removal of junk, i.e., paper clips, wire, rope, and staples. Exceptions to the cleaning operation are the high grades of paper stock or pulp substitutes such as hard white envelope cuttings, which may go directly to the

82

pulper without cleaning. This fact needs to be borne in mind when institutional and economic comparisons are made between virgin pulp and recycling processes.

Concluding Comment

This brief overview of paper as a good that man fabricates from a natural substance has discussed the kinds and uses of paper, the residuals generated by its production and consumption, and how paper residuals may be reused before they are returned to the environment. The prevailing conditions and the possible alternate uses of paper are governed by economic and institutional considerations. The next step is a detailed examination of the handling of newspaper and corrugated container residuals in the DC/SMSA.

5

The Generation, Disposal, and Recycling of Newspaper and Corrugated Container Residuals in Metropolitan Washington

Introduction

The extent to which recycling of solid residuals can be increased in municipal areas depends on the supply of and demand for the residuals as raw material. The cost of residuals supply is a function not only of the labor and equipment costs of aggregating residuals as raw material, but also the costs of disposing of the residuals if they are not recycled. Under what circumstance is the value of the recovered residual high enough as a raw material to equal or exceed the cost of recovering the residual? Does the diversion of the residual from the solid waste stream result in any noticeable reduction in the priced costs of managing mixed solid residuals and the unpriced costs of damages resulting from discard to the environment? In an effort to develop the proper context for asking these questions, and also to discover some of the factors involved in the answer, the flow of

85

newspaper and corrugated container residuals has been traced through the economy of the Washington metropolitan area.

Newspapers and corrugated containers are logical residuals for this study because they are the two largest homogeneous fractions of municipal mixed solid residuals. As such, their sources, quantities, and dispositions are the easiest to trace, and their diversion in any substantial amount should also yield the greatest impact on the solid residuals handling and disposal system.

Furthermore, Washington is a paper town. Paper is the major tangible output of the federal government; printing is the number two industry in the metropolitan area after government; three[1] major daily newspapers are printed there; and Washington is a major wholesale distributing center for the Baltimore–Richmond corridor, which implies the presence of a large quantity of corrugated container residuals.

Newsprint Products and Newspaper Residuals in the DC/SMSA

Figure 4 shows the flows of newsprint and newspapers in the standard metropolitan statistical area of the District of Columbia. Newspapers are a relatively easy product on which to obtain actual input data for a materials balance, because of the tally points and records that are developed in the manufacture and distribution of newspapers. As in any production system, records of quantity of material used must be kept to determine costs and efficiencies. In addition, data on the distribution of a particular newspaper are available from the audit report of that paper prepared by the Audit Bureau of Circulations of Chicago. Audits of circulation are made and reported periodically by the bureau for the benefit of newspaper advertisers. Thus data were available on the tonnage of newsprint run on the newspaper plant presses, and on the distribution of this tonnage within and beyond the surrounding area.

The tonnage attributable to inserts and sections printed outside the newspaper plant and folded into the newspaper editions for distribution, such as magazine sections, book review sections, advertising inserts and comic strips, was not calculated with the same degree of accuracy because figures were not readily obtainable on the tonnage of these papers run on

---

[1] The number of daily newspapers shrank from three to two during the time of the study reported here.

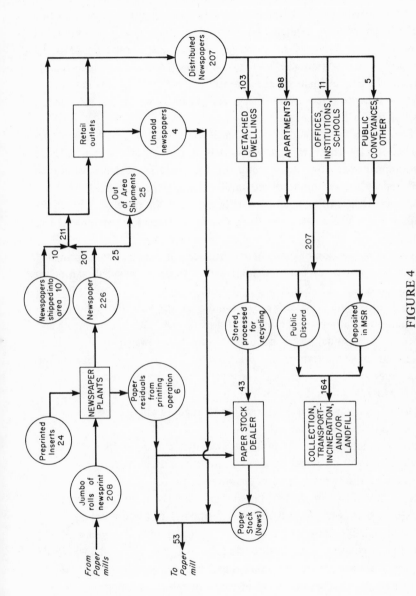

FIGURE 4

Flow of newsprint and newspaper through the DC/SMSA, 1969. (All figures in 1,000 tons.)

outside presses; however, some degree of accuracy was achieved by obtaining the total number of pages of inserts for a given period and combining this figure with sampled weights. With the cooperation of outside newspapers, figures were calculated on the total weight of newspapers imported into the area. Most of the tonnage of newspapers imported into the DC/SMSA was attributable to the *New York Times.* Other high-volume papers shipped in included the *Wall Street Journal.* This was not considered an area paper even though it is printed in Silver Spring, Maryland, a part of the DC/SMSA.

The fourth category of newspaper input for the DC/SMSA is papers carried in by travelers. Since this was a very elusive datum, it was determined that the requirements of the materials balance would be satisfied by the assumption that travelers leaving the DC/SMSA carry out the same weight of newspapers carried by incoming travelers. The inputs and outputs of the newsprint flows shown in figure 4 are summarized in table 10.

The output section of the materials balance is made up of three principal categories: shipments out, newsprint and newspapers recycled, and newspapers disposed of. Shipments out were calculated by associating the weights of weekday, Saturday, and Sunday papers with the Audit Bureau circulation data beyond the boundaries of the DC/SMSA. Converting and distribution residuals were calculated from the newspapers' records of shipments of scrap, returns, and overissue news to paper stock dealers. The remainder of the output side consists of consumption residuals, which totaled 207,000 tons. Forty-three thousand tons of this total are estimated to have been recycled, and 164,000 are estimated to have been disposed of by incineration and/or landfill.

The estimate of 43,000 tons of newspaper consumption residuals recycled in 1969 is based on a survey of the thirteen retail paper stock dealers and the two wholesalers operating in the DC/SMSA. The amounts handled by different dealers varied tremendously from less than 1,000 tons a year to well over 10,000 tons. Estimates of the operations of the thirteen retail dealers in consumption residuals plus the tallies of the tonnage of converting and distribution residuals handled by the wholesale paper stock dealers totaled 53,000 tons. One of the wholesalers, asked to estimate the Washington output of newspapers for recycling on the basis of his knowledge of the local market, quoted a figure of 1,000 tons a week. The recycling estimate of 53,000 tons a year, made without benefit of access to accounting records, appears to be reasonably conservative.

TABLE 10
Materials Balance for Newsprint in Washington for 1969
(thousands of tons)

Material	Quantity
Inputs	
Tonnage run on in-house presses: three major D.C. dailies, plus estimated tonnage of small area papers.	208
Inserts and sections printed on outside presses	24
Newspapers shipped in	10
Newspapers carried in by travelers	?
Total inputs	242
Outputs	
Shipment outside the DC/SMSA	25
Carried out by travelers	?
Recycled (paper stock shipments to mills)	
Converting residuals 6	
Distribution residuals 4	
Consumption residuals 43	53
Consumption residuals disposed of	164
Total outputs	242

Sources: Data and estimates compiled from field observations. An assumption has been made in this balance that 100 percent of distribution and converting residuals are recycled.

The estimate of 164,000 tons of newspaper residuals disposed of by incineration or landfill is derived by deducting recycled tonnage from total input. This estimate does not allow for any additions to inventory, which are so small as to be insignificant.

Factors Affecting Recycling

The significance of the qualities of concentrated mass, homogeneity, and uniformity of contamination in determining whether or not residuals are recycled has already been noted. It is these qualities which to a considerable extent determined in 1969 that 43,000 tons of consumption residuals were recycled and 164,000 tons were disposed of, rather than some other combination of figures. Maintenance of homogeneity in newspaper residuals is aided by the capability of visually identifying the product. Anything that looks like a newspaper is made of newsprint—a combination of groundwood and chemical pulps. Concentration of mass and uniformity of contamination are strongly influenced by the circumstances in which the residuals are generated, as will be seen later. Figure 4 indicates the point of generation and hence the extent of the dispersion of newspaper consumption residuals. The 207,000 tons of newspapers reaching consumers in the DC/SMSA finished their useful life in dwelling units, institutions, commercial establishments, public conveyances, or other public areas. Circulation data do not indicate where newspapers are read. There is information in the Audit Bureau reports on the number of copies delivered to residences in certain zones; however, it is not possible to distinguish, for example, between newspapers delivered to office buildings and newspapers delivered to retail outlets. It is logical to assume that the preponderance of newspapers sold in retail outlets, especially on Sundays, will be consumed and become residuals in residences. An estimate that 92 percent of all newspapers are read in residences has been noted. For the remainder, an arbitrary allocation was made of 70 percent to offices, institutions, and schools, and 30 percent to public conveyances and other public places. Applying these proportions to the 207,000 tons yields estimates of 181,000 tons of newspaper residuals generated in residences; 11,000 tons in offices, institutions, and schools; and 5,000 tons on public conveyances and in other public places.

The household census for the DC/SMSA in 1969 shows there were 506,000 single-family dwellings and 425,000 multiple-family dwellings. These comprise, respectively, 54 and 46 percent of all dwellings. Applying these percentages to the estimated newspaper consumption residuals in the DC/SMSA yields 103,000 tons of newspapers available from single-family dwellings and 88,000 tons from multiple-family dwellings and apartment houses. In the absence of housing data according to income levels, this allocation assumes uniform distribution of newspapers to all residences for theoretical purposes only. It was pointed out above that households with persons of higher income and higher education are more apt to have newspapers than households with persons of lower income and less education. It is also obvious that households grouped together in apartment buildings will yield a larger number of newspaper residuals for a given expenditure of collection effort than the same number of households in detached residences. These facts are important to the development of any municipal plans for separate collection of newspapers. It should also be

90

kept in mind that generally municipalities collect residuals only from one- or two- or three-family buildings, with private haulers collecting from apartment houses.

Potential for Separate Handling of Newspaper Residuals

Uniformity of contamination is the critical factor in recycling newspaper residuals. If newspaper residuals can be kept separate from time of generation through on-site handling, processing, and storage, and off-site collection, then they may be used to produce newsprint or other paper products. If they cannot be kept separate and are stored and collected with other solid residuals, the potential for economic recycling is substantially reduced. An objection frequently made to household separation for recycling is that householders will not separate newspapers voluntarily, and not very well, even under regulation requiring separation. This point of view stimulates the question of how householders handle their newspapers under conditions where there is no municipal ordinance requiring separation and no centralized recycling program other than sporadic drives by Boy Scouts or churches. To obtain some data on this question, two collecting forays for newspaper residuals were made in Arlington County, a component of the DC/SMSA. One was in a collection route in the census tract having the highest median income in the county, the other in one of the lowest median income areas.[2] Both areas consisted primarily of single-family dwellings, and collections were made only from structures collected by the county, that is, three families or less.

In these two samplings six different methods of on-site handling and storing of newspaper residuals were identified, more than one of which could occur at a single site:

(1) piled loose, usually on the ground beside the containers, but separated from other solid residuals;

(2) bundled separately—tied with string or women's stockings; in grocery sacks; in cartons;

(3) placed in a separate can, usually vertically, indicating some period of accumulation, i.e., a week or two since last collection. The newspapers apparently had been placed in the can all at once, or a separate can was

[2] These forays were made possible by the cooperation of Harry Doe, then Director of Utilities of Arlington County.

used for accumulating during the time period. If the former was true, then there had been an intermediate storage point between consumption and the trash can;

(4) discrete, folded flat, and placed in varying amounts on top of one or more containers;

(5) discrete, folded flat in small amounts but intermingled with other residuals, indicating they had been put out daily, more or less with other residuals;

(6) crumpled and mixed with other residuals so that discrete amounts could not readily be removed. Sometimes reflected use in household, i.e., to catch paint drippings, line pet boxes, and so forth. A number of trash setups, perhaps 25–30 percent, were in tied plastic bags, preventing any examination of the MSR for discrete papers. It is probable that newspapers in the plastic bags were mixed with other residuals rather than discrete.

The results were unambiguous for both income areas—household separation of newspapers, available for separate collection, is readily achievable. In fact, it tends to prevail in the absence of any outside stimulus. Probably the British government of 1711 which helped to determine the page size of the present-day newspaper is the primary influence on the on-site storage and processing of this particular residual following residential consumption. The size of newspaper pages makes papers much easier to handle in a flat, folded condition in which they occupy less space. By the same token, they are extremely wasteful of space when placed in the usual 20- or 30-gallon household trash can.

Several other pertinent data were noted on these observation trips. First, newspapers may be put out separately because the householders are consciously or subconsciously hopeful that they will be recycled. On both forays there were a number of householders who, seeing newspaper collectors working along the street, came to their doors to ask if the collectors would like the newspapers accumulated in their basements.

Second, a major difference between the high income and the low income areas is the absence of sink garbage grinders in the low income area. The food scraps put out to be collected with other solid residuals in the low income area had a high moisture content which tended to migrate to surrounding newspapers, diminishing their quality for recycling. Also, newspapers were used to wrap garbage, increasing the proportion of mixed and crumpled newspapers noted in the setups.

Third, newspaper residuals comprised about 10 percent of the total MSR collected in the upper income area, and about 5 percent in the lower income area. Both figures discount the proportion of newspaper residuals to total MSR generated because only discrete accumulations were collected. This is particularly true of the upper income area, where there was more evidence of newspaper residuals stored in garages or basements awaiting special collection drives than in the lower income area.

There are at least four ways in which newspapers in single-family residences are diverted from municipal collection of mixed solid residuals to recycling. The most pervasive is the philanthropic organization project for raising money. Schools, churches, Boy and Girl Scout troops are the common examples. Some churches and schools make this a year-round activity, while for other groups it tends to be a seasonal drive. In these cases the groups collect from the individual residences.

A second method involves the "gypsy," the scavenger who goes along and picks paper off the trash setups ahead of the municipal collection trucks. (This was the procedure used in the forays discussed above.) This operation comes into existence when paper stock dealers are paying higher prices for paper and goes out of existence when the price goes down. The 30 cents per hundred pounds that prevailed during most of this study was not attractive to gypsies. There would be a good deal more of such activity if the price were 60 cents per hundredweight.

A third way involves the individual citizen who, for whatever reason, will accumulate his papers and deliver them to the paper stock dealer.

The final way involves the paper stock dealer providing a storage container at a central location, such as a churchyard or supermarket parking lot, to which individual citizens bring their newspaper residuals. The dealer periodically collects the papers deposited in the storage container.

Availability of Newspaper Residuals in Apartment Houses

The on-site handling, processing, and storage of newspaper residuals generated in apartment house complexes follow various patterns. There is a difference between the patterns in garden and high-rise apartments, reflecting the horizontal and vertical distribution of dwelling units. A common routine for a garden apartment is for the householder to carry his refuse to the trash room serving his particular section. Here the residuals are stored in the conventional 30-gallon cylindrical container which must be tipped manually into a collection truck, normally a rear-loading compactor. The residuals may be collected from the trash rooms daily or every

93

other day. A fairly recent modification of this practice, with adverse impact on aesthetics, is the placement of large portable containers on castors on the edge of driveways in the garden apartment complex, where they are tipped by a front-end loading refuse truck. This is a less expensive form of storage and collection than trash rooms.

The potential for recycling newspapers from garden apartments is about the same as for recycling from individual residences. Because of lack of storage space in most cases there is probably less possibility of accumulating papers for a special drive. However, with the presence of a number of households aggregating residuals in one location for a daily or every-other-day pickup, the potential for separate storage in trash rooms and containers is theoretically good. One of the characteristics of garden apartments with dispersed trash rooms is the hazard of children playing with matches and setting fires. For this reason, a number of garden apartment managers indicated reluctance to provide for separate pickup of newspaper residuals. This is not necessarily a rational concern, because flat, folded newspapers are not highly flammable. Nonetheless, the concern is present.

Unlike the garden apartments, of which only two in the Arlington survey kept newspaper residuals separate, a number of high-rise apartments provided for separate handling, processing, and storing of newspaper residuals. While some of this activity was undoubtedly due to the transition from incineration to no incineration, it is also true that it appeared to be economically beneficial to keep newspapers separate for recycling. The diversion of newspaper residuals to recycling reduces the amount of solid residuals to be hauled to the incinerator or the landfill, which may or may not result in a realizable saving to the apartment house owners. If these costs are computed on the basis of the weight and volume of the residuals, as they may be if the apartment management hauls its own refuse, there may be a saving. If the apartment house owners hire a trash hauler on a fee fixed per dwelling unit, there will probably not be any savings realized. Most trash hauling contracts today are made on the basis of a fixed monthly fee for each dwelling unit in the apartment house. Only if the fee were reduced to reflect a lower quantity per unit would savings be realized.

Three different kinds of arrangements for segregated storage and collection of newspaper residuals were found in a survey of apartment houses in Arlington that were five stories or over. In the first arrangement, whether or not trash chutes were used for MSR, newspapers were taken by hand or by cart from each apartment to the central trash room of the high-rise by the tenant. In some instances tenants left them in the common utility room on the floor on which they lived, and porters trucked them to the elevator and from the elevator to the residuals storage area. In other cases tenants carried them separately to the residuals storage area. In some instances they were stacked (not bundled) in the trash room. A gypsy collector, generally moonlighting, hauled the papers out three or four times a week. He obtained the papers at no cost and sold them to a local paper stock dealer. One such operator appeared to be collecting 10 to 12 tons a month.

The second and third arrangements both involved the placement, by a paper stock dealer, of a 10- or 12-cubic yard portable bin or "dumpster" in the apartment house service area. Residents placed their papers in the hallway. Apartment house porters toted the newspapers to the portable bin, which was picked up once a week by the paper stock dealer. In the second arrangement, the paper stock dealer paid for the newspapers he collected; in the third arrangement, he did not. This difference appeared to depend on the cleanliness of the newspaper residuals. If some general refuse was thrown in the bins with the newspapers, it lowered the value to the paper stock dealer. One particular arrangement of a portable bin in an 1,100-unit high-rise apartment complex yielded a total of 310 tons of newspapers in the course of a year. In these arrangements newspapers were not incinerated even if other MSR were because of the combustion difficulties apartment houses have with newspapers.

The effect of site design on the potential for additional recycling of newspapers was apparent in the surveys made for this study, both in the case of single-family residences and multiple dwellings. Provision of convenient storage space in detached residences facilitates separate handling. The relationship of garbage grinders to residuals storage was clearly shown. The pattern of urban living, shifting from detached single-family dwellings to high-rise apartments, increases the concentrated mass of newspaper residuals available for pickup. In Arlington County, single housing units increased by 3 percent from 1960 to 1970, while multihousing units increased by 47 percent. In 1960, 50.2 percent of all dwelling units were single; in 1970 the percentage of single dwelling units had declined to 41.1 percent. Thus the long-range planning in the DC/SMSA and similar metropolitan areas for solid residuals handling should take into account the configuration of the residential generating sources.

The on-site generation, storage, and processing of the estimated 191,000 tons of newspaper residuals in the residences of the DC/SMSA thus varies with income, the presence or absence of kitchen garbage grinders in both single- and multiple-family dwellings, the extent to which the dwellings are detached, grouped horizontally, or grouped vertically, and the physical design of the residential structures. The estimated 11,000 tons of newspaper residuals to be found in places of work and the estimated 5,000 tons of newspaper residuals discarded on buses and in other public places were not considered good candidates for recycling. In the case of offices, there are more important paper fractions to be recovered, both in terms of relative volume and relative value as raw materials. In the case of public discard, the potential for recovery of an uncontaminated fraction of a significant size appeared negligible.

Potential for Increased Recycling of
Newspaper Residuals in the DC/SMSA

What portion of the 164,000 tons of newspaper residuals that were discarded in 1969 can be economically recycled and under what conditions? On the basis of the total proportion of detached and semidetached housing units in the DC/SMSA, it was estimated above that 103,000 tons of the 191,000 tons distributed to households went to single or duplex units. It is estimated that 25 percent of the 103,000 tons (26,000 tons) was not available because it had been diverted to other uses, such as wrapping garbage, miscellaneous household clean-up, or was handled by householders not responsive to public programs. It was also estimated that of the 43,000 tons of newspaper consumption residuals recycled in 1969, 30,000 tons came from single-family residences. This leaves a balance of 47,000 tons of newspaper residuals "readily available" from single-family residences in the DC/SMSA. The 70 percent figure, 30,000 tons, is estimated on the basis that, while the apartment house dweller is not in a good position to handle newspaper residuals for recycling, the single-family householder is, because of storage space and easy access to transportation.

Of the 88,000 tons of newspapers residuals allocated to apartment houses, it is estimated that 25 percent or 22,000 tons will not be available, primarily from garden apartments which tend to lack garbage grinders and easily accessible storage. It is estimated that 30 percent (13,000 tons) of the 43,000 tons of newspaper residuals recycled came from the 88,000 tons, leaving as "readily available" 53,000 tons from apartment houses.

96

The readily available tonnage of 47,000 tons from single-family residences and 53,000 tons from apartment houses yielded an estimated 100,000 additional tons of newspaper residuals available for economically efficient recycling. The effect of this additional recycling of newspaper residuals on solid residuals management costs is discussed in the next chapter.

Generation of Corrugated Container Residuals in the DC/SMSA

The calculation of a materials balance for corrugated containers in the DC/SMSA in 1969 was not determined as readily as the balance for newspapers. There are central tally points from which data on the quantity of newsprint used can readily be determined. There are no similar local central points from which quantities of corrugated cartons in the area can be derived. In the absence thereof, the procedure was to develop per capita wholesale trade factors for the nation and for the DC/SMSA, from which a per capita consumption of corrugated containers in the District was derived. The result was an annual per capita District consumption of 89.5 pounds of corrugated containers compared to a national per capita average of 132 pounds. The DC/SMSA per capita consumption multiplied by the population in the area yielded a figure of 130,000 tons of corrugated container residuals in the District in 1969. This appears a reasonable figure when the wholesale activity of the DC/SMSA is compared with other SMSAs, and the absence of assembly manufacturing activity in the DC/SMSA is noted.

A further complicating factor is that the economic data indicating the consumption of corrugated containers in the DC/SMSA does not necessarily imply the continuing presence of the consumption residual in the DC/SMSA. For example, one of the largest food chains in the area is served by a distribution center in Baltimore. The containers used for goods sold in the DC/SMSA are trucked back to the distribution center outside the area. Two chains with distribution centers in the DC/SMSA recover containers used for goods sold in Pennsylvania, Delaware, and Richmond, Virginia. For this reason calculations of availability of corrugated container residuals on the basis of wholesale and retail sales in the area were distorted.

A survey of paper stock dealers in the area indicated a processing of 13,000 tons of corrugated container board residuals in 1969. A survey of the supermarket chains indicated the recovery of about 33,000 tons for a total recovery of 46,000 tons of the 130,000 tons estimated to be generated in the area. How much more can economically be recycled? This involves a detailed examination of circumstances of the flows which are shown in figure 5.

Goods packed in corrugated containers entering the DC/SMSA may have an intermediate stop in a warehouse, or they may go directly to the site of consumption. Figure 5 shows goods traveling to office buildings, institutions, and restaurants as well as to retail stores. The greatest tonnage of corrugated container residuals, by far, is generated by retail food stores. The 33,000 tons shown in figure 5 as coming from warehouse and wholesale activities came from just two supermarket chains where stores generated from 6 to 12 tons of corrugated container residuals in one week. Supermarket chains have an ideal arrangement for aggregating large masses of homogeneous corrugated container residuals because the large trailers which deliver goods from the distribution center to the retail outlet are available at literally no cost to return the used containers from the stores to the center. In one system observed in the DC/SMSA, the stores baled the residuals in relatively small bales weighing from 175 to 250 pounds. This on-site processing operation reduced space required for storage of residuals and facilitated such on-site handling as mechanized movement from the storage area to the truck loading dock and loading on trailers. On return to the distribution center, where there is a building designed especially for handling and processing corrugated container residuals, the bales are broken open, and the corrugated residuals are rebaled into bales averaging 1,800 pounds and measuring 60 cubic feet. These bales are stored and loaded into freight cars by fork lift trucks. The conveyors and baling equipment required for this operation represent a capital investment on the order of $150,000.

This tonnage was shipped directly to paper mills from the supermarket distribution centers. It did not move through the plants of paper stock dealers; however, paper stock dealers did act as brokers for the transactions between the centers and the mills and in some cases acted as carriers for the container board when truck shipment was used.

A survey of the thirteen paper stock dealers in the DC/SMSA indicated that seven of them had bought and sold corrugated container residuals in 1969 ranging from a low of 200 tons to a high of 5,200 tons for a total of 13,000 tons.

This quantity came from retail food stores, not organized as the supermarket chains were, from department stores, liquor stores, and drug stores.

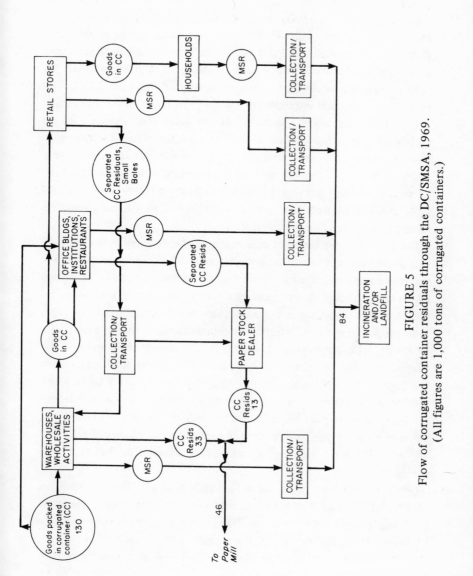

FIGURE 5

Flow of corrugated container residuals through the DC/SMSA, 1969.
(All figures are 1,000 tons of corrugated containers.)

The gigantic establishment of the federal government in the District of Columbia did not show any activity in recovering corrugated container residuals in 1969.

The estimate of 130,000 tons of corrugated residuals entering the DC/SMSA in 1969, with 46,000 tons recovered, leaves 84,000 tons discarded to incineration and/or landfill. The flow of corrugated going into noncommercial activities—households, institutions, and so forth—on the basis of national percentages is estimated to be about 12,000 tons. This flow is so diverse and intermittent that the mass for economic recycling is not present; this amount therefore is not considered available for recycling. It also estimated that 12,000 tons of the remaining corrugated flow in commercial establishments is not available because of highly dispersed locations, low volume, and high contamination. Thus, the total of readily available corrugated container residuals is estimated to be about (130–24) 106,000 tons per year of which 46,000 are currently recycled, leaving 60,000 tons as readily available.

The major sources for additional recovery of corrugated container residuals appear to be retail stores, particularly food and beverage stores. The volume of nonretail, nonmanufacturing institutional generation is not apt to be great enough to warrant the special arrangements required for handling and processing corrugated container board. A study of paper residuals management in offices[3] has estimated that a county government of 43 departments employing 10,000 people generated 48 tons of corrugated container residuals in a year's time. This compares with a recovery (perhaps 65 percent of generation) figure of 84 tons for a single supermarket in a year. The possibility of getting less than a ton a week from a scattering of county buildings compared with the possibility of collecting 1½ tons a week from a single store illustrates the economies of scale that are present for the recovery of corrugated residuals in the retail food stores.

Of the 60,000 tons considered to be readily available, it is estimated that 60 percent or about 36,000 tons are available from high-volume generators who would make bales weighing 200 pounds or more which would be mechanically loaded on trucks from loading dock levels. The balance, 24,000 tons, is considered to come from low-volume generators who would make bales of about 45 pounds which can be manually loaded on

[3] Charles N. Ehler and E. J. Beltrani, "Paper Residuals Management in Offices," in press, U.S. Environmental Protection Agency.

trucks at locations not having truck level loading docks. The collection of high-volume corrugated residuals which can be mechanically loaded in sizable increments will be less expensive than the collection of the low-volume corrugated residuals.

Summary

It is apparent that there has been little organized attention focused on on-site handling, processing, and storage of both newspaper residuals and corrugated container residuals. Despite this lack of attention and the absence of formal governmental incentives, respectable fractions of these residuals generated in households and commercial establishments have been recycled within current economic constraints and incentives. Where newspaper residuals occur so that they can easily be collected in quantity and in clean condition, a significant amount of recycling is already taking place. An analysis of the factors which contribute to this level of recycling should indicate the potential for increasing the level.

Similar considerations apply to the potential for recycling corrugated container residuals in the area. Large-volume sources of relatively uncontaminated material, such as supermarkets and other large stores, are the logical candidates for economical recycling. The necessity for volume reduction to achieve efficient handling simply for disposal increases the possibility that significant quantities can be recovered from smaller generators such as independent stores and restaurants. However, corrugated container board residuals from households are generally so dispersed and of such low volume that they do not have a sufficient concentration of mass to offer good recycling potential.

With newsprint comprising 9.9 of the 58.6 million tons of paper consumed in the United States in 1969 and corrugated container board comprising 14 million tons, for a combined total of about 24 of the 59 million tons, systems that provide for increased recycling of these two categories of paper may have significant impacts on solid residuals handling and disposal. This is explored in the next chapter.

6

Effects of Recycling
Newspapers and Corrugated
Containers on Solid
Residuals Management Costs
in the DC/SMSA

Introduction

The effects of recycling additional quantities of newspaper and corrugated container residuals on the costs of solid residuals management in the DC/SMSA can be estimated by aggregating the quantities of residuals collected and disposed of in the various jurisdictions of the metropolitan area and computing the operation costs for the systems as they operated in 1969–70. This model of the actual system can then be modified by substitutions of quantities of newspapers and corrugated containers diverted to recycling, resulting in varying system costs. This chapter examines the actual model and the results of hypothetical changes in quantities recycled.

The governments of the District of Columbia, Montgomery County, Prince George's County, Alexandria city, Arlington County, Fairfax city,

TABLE 11

Solid Residuals Quantities and Costs, 1969–70 for DC/SMSA[a]

Quantities and costs	District of Columbia	Montgomery Co.	Prince George's Co.	Alexandria city	Arlington Co.	Fairfax city	Fairfax Co.	Falls Church city
MSR collected (10³ tons)	599	300	338	73.6	103	12.3	208	6.0
Unit collection costs ($/ton)	25	25	25	15	20	20	20	20
Collection costs (10⁶ $)	14.97	7.50	8.44	1.10	2.07	0.25	4.16	0.12
MSR incinerated (10³ tons)	322	300	0	73.6	104	0	0	0
Unit incineration costs ($/ton)	8	8	—	8	8	—	—	—
Incineration costs (10⁶ $)	2.58	2.40	0	0.59	0.83	0	0	0
MSR directly land-filled (10³ tons)	277	0	338	0	0	12.3	208	6.0

Unit direct landfill costs ($/ton)	3	3	3	—	—	3	—	3
Direct landfill costs (10^6 $)	0.02	0.62	0.04	0	0	1.01	0	0.83
Incinerator residue landfilled (10^3 tons)	0	0	0	49.7	22.7	0	150	184
Incinerator residue landfill costs ($/ton)	—	—	—	2	2	—	2	2
Incinerator residue landfill costs (10^6 $)	0	0	0	0.09	0.05	0	0.30	0.37
Total landfill costs (10^6 $)	0.02	0.62	0.04	0.09	0.05	1.01	0.30	1.20
Total solid residuals management costs (10^6 $)	0.14	4.78	0.29	2.99	1.74	9.45	10.20	18.75

[a]Total annual cost for the DC/SMSA is 48.34 × 10^6 $.

Fairfax County, and Falls Church city (the DC/SMSA with the exception of the rural counties Loudoun and Prince William) reported disposing of quantities of mixed solid residuals totaling about 1.64 million tons in 1969-70. The quantities and operation costs for each jurisdiction are set forth in table 11, which indicates a total cost of about 48 million dollars.

The accuracy of the figures in table 11 is limited by jurisdictional variations in measurement techniques and record-keeping methods. For some jurisdictions the reported weights are the totals of truckload weights tallied at incinerator and landfill scales. Other jurisdictions not having scales at the disposal sites derived their annual tonnage totals by converting tallies of truck loads on an arbitrary basis of 150 or 180 pounds for a loose yard of refuse and 500 or 600 pounds for a compacted yard of refuse. In still other cases, total solid residuals reported incinerated and landfilled were computed by multiplying an average per capita collection of MSR by the total population in the jurisdiction.

The collection of the 1.64 million tons of mixed solid residuals reported in table 11 was carried out by (a) departments of the municipal administrations, (b) commercial haulers under contract to the municipal administrations, (c) commercial trash haulers under contract to private firms (there are 115 companies in the SMSA listed in the classified telephone directory), and (d) the individual generators themselves. The unit collection costs presented in the table are the costs reported by the various jurisdictions for that portion of MSR which they collected or contracted to have collected. Generally this portion is generated by private residences only with no more than two, or, in some jurisdictions, three dwelling units. The costs of commercial haulers were not available for these calculations. While commercial and private haulers handle larger quantities than municipal departments, for purposes of this analysis the costs of collection reported by the public agencies were assumed to apply to all tonnage collected in the various localities. In some jurisdictions the costs were modified to gain comparability with such elements as administrative and fixed overhead, depreciation, and debt services. Unit collection costs for each jurisdiction were rounded for consistency.

It was possible to obtain tallied weights of incinerator residue in only three of the four jurisdictions having incinerators. In the fourth jurisdiction, residue was estimated at 50 percent of input. On this basis, total residue varied only 1 percent from being one-half the weight of all mixed solid residuals incinerated in the SMSA and for purposes of this discussion is considered to be 50 percent of input by weight.

It was not possible to determine from incineration costs reported that these costs covered all maintenance and capital costs, including costs of land. The most carefully analyzed cost was $8 per ton reported by one jurisdiction. Because of the explicitness of the elements which made it up, and because of its apparent completeness, this cost was adopted for all the jurisdictions. These data relate to costs *prior* to the increased stringency of air pollution controls.

Reported costs per ton for direct landfill ranged by jurisdiction from $1.11 to $2.50, where the cost was segregated from other disposal costs. These appeared to cover direct labor and operation and maintenance of equipment used at the landfill. It is not certain that they covered any depreciation of equipment, and it is quite certain they did not cover debt service. Since the time these data were gathered, two jurisdictions have imposed charges of $5 per ton for landfill or incineration on commercial and private haulers. For the initial computation of total annual costs, a landfill cost of $3 per ton was adopted for all jurisdictions to include fixed charges for capital and land plus operating and maintenance costs.

To reflect actual operating conditions in the solid residuals management system, a lower cost, i.e., $2 per ton, has been assigned to the landfilling of incinerator residue than to the direct landfilling of mixed solid residuals. Incinerator residue landfills at a density in excess of 1,000 pounds per cubic yard without compaction, while unburned mixed solid residuals require thorough compacting to achieve a density of 700 to 800 pounds per cubic yard. Furthermore, incinerator residue travels to the landfill at a natural density of 500 to 600 pounds per cubic yard.

In summary, the DC/SMSA disposed of about 1.64 million tons of MSR, about 50 percent by incineration and landfill of incineration residue and about 50 percent by direct landfill, at a total estimated cost of about 48 million dollars.

Integration of Residuals Flows to Recycle and to Final Disposal

The total residuals handling system of the DC/SMSA includes quantities of residuals recycled as well as quantities of residuals disposed of.

Within the quantities of residuals finally landfilled is some portion which, under different market conditions or different institutional arrangements, might economically be recycled. In chapter 5 the flows of 217,000 tons[1] of newsprint and newspapers and 130,000 tons of used corrugated containers in the DC/SMSA were analyzed. These flows are integrated with the total solid residuals flow in figure 6 to indicate the total actual flows in 1969–70 and the variations that would result from additional recycling.

The 1,392,000 tons of all other solid residuals shown on the bottom flow line to processing and landfill are augmented by the 164,000 tons of newspaper residuals and the 84,000 tons of corrugated container residuals not recycled in 1969 to make the total of 1.64 million tons of mixed solid residuals discarded. The 53,000 tons of newspapers and the 46,000 tons of used corrugated containers recycled in 1969 are excluded from the 1.64 million tons of MSR collected for disposal.

There is some quantity of other types of residuals which are currently recycled in the DC/SMSA. These are shown in the diagram but without amounts estimated. These residuals include the recovered and recoverable metals, glass, and textiles. As with the residuals explicitly analyzed, the quantity recycled is not included in the 1.64 million tons; the quantity not recycled is included.

In this chapter the total annual costs of three different patterns of the flow of solid residuals through the solid residuals management system[2] of the DC/SMSA are compared. For the sake of editorial convenience the different patterns are designated solid residuals management system (SRMS) I, II, and III. The first one is essentially the system as it operated in 1969–70. It is the base line case for comparison with the other two systems and for sensitivity analysis. The second system assumes additional recycling of newspaper residuals. The third system assumes additional recycling of both newspaper residuals and corrugated container residuals.

Solid Residuals Management System I

A simplified and consolidated flow system of all mixed solid residuals collected and discharged to municipal incinerators and landfills in the DC/SMSA in 1969–70 is shown in figure 7. An alternate flow line from collection to landfill, bypassing incineration, is shown for half of the

---

[1] The 25,000 tons of newspapers circulated outside of the DC/SMSA are not included in this quantity.

[2] *System* as used here denotes the composite of all arrangements for handling solid residuals in all the jurisdictions of the DC/SMSA.

residuals. This, in fact, is the situation that prevailed in the DC/SMSA. Obvious modifications of the system which can be calculated to check system response to various operational changes are to hypothesize that all residuals go directly to landfill, without incineration, or alternately, that all residuals go to incineration with only the residue landfilled. The calculation of the annual cost of the flows in figure 7 is expressed by an equation shown in the appendix to this chapter.

The total costs for each operation and the total annual costs for the system are shown in table 12. The various unit operational costs shown for each jurisdiction in table 11 have been compressed into weighted averages for the entire SMSA, and the quantities collected, incinerated, and land-filled have been aggregated to give single quantities for each operation. Six different system variations are shown in table 12.

The variations are divided into three groupings, I and I' in which MSR are half incinerated and half directly landfilled without incineration; IA and IA' in which all MSR are landfilled without any incineration; and IB and IB', which show all MSR are incinerated and the incinerator residue is landfilled.

Within each grouping the disposal cost is varied from a low value to a high value. Thus in I and IB the unit cost of incineration is a low value, $8 per ton with $2 per ton for landfilling the residue. In I' and IB' the unit cost of incineration is posited at $12 per ton with a charge of $4 per ton for landfilling the incineration residue.

In system I and IA the system is costed out with a low unit cost of $3 per ton for direct landfilling of MSR. In I' and IA' the high cost of direct landfilling, $5 per ton, is posited.

Comparison of these various operational patterns indicates that the least expensive system is SRMS-IA in which all MSR are landfilled directly at the low landfill cost of $3 per ton, and the most expensive system is IB' in which all MSR are incinerated and the residue landfilled at higher unit costs. It should be noted that higher unit costs for incineration may be anticipated, reflecting the additional technology required to meet higher air quality standards being imposed in Virginia, Maryland, and the District of Columbia.

Solid Residuals Management System II

In the system depicted in figure 8 the total annual cost of solid residuals management is analyzed as it is affected by recycling additional newspaper residuals, with varying arrangements for incineration and landfill for the

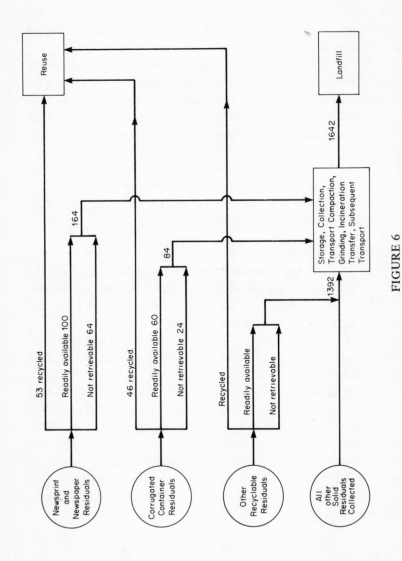

FIGURE 6

Flows of residuals to reuse or landfill in the DC/SMSA, 1969. (All figures in 1,000 tons.)

Data do not include abandoned vehicles, demolition wastes, or the residuals of clearing sites for construction. These residuals are usually disposed of to private landfills. The 1.64 million tons does not include quantities estimated for Loudoun and Prince William Counties, which, as primarily rural areas, have residuals disposal characteristics unlike the rest of the SMSA. The quantities are relatively small and have negligible effect on subsequent calculations. The dividing line between "currently recycled" and "readily available" is not a firm and fixed boundary. It varies with the demand for secondary materials and the costs of collecting and processing residuals. It is simply a reflection of the market at the time considered. "Readily available" could become "currently recycled" with modifications of the market. Likewise the dividing line between "readily available" and "not retrievable" is not capable of being fixed, depending as it does on changing economic, technological, and political factors.

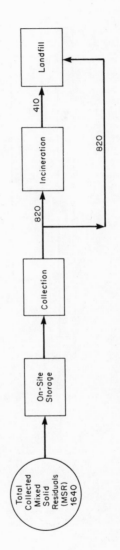

FIGURE 7

Solid residuals management system I: flow of MSR through the system. (All figures in 1,000 tons.)

111

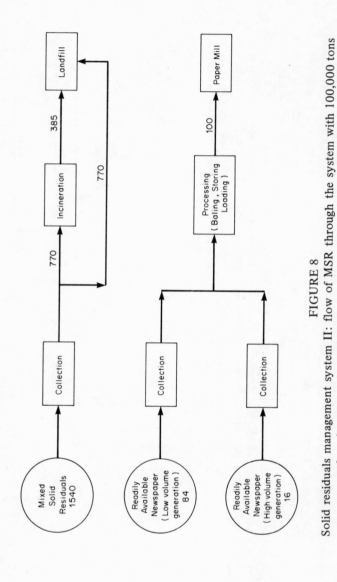

FIGURE 8

Solid residuals management system II: flow of MSR through the system with 100,000 tons of used newspapers diverted to recycling. (All figures in 1,000 tons.)

112

TABLE 12
Total Annual Costs for Solid Residuals Management System I
and Variations[a]

	MSR 50% incineration 50% landfilled		MSR 100% landfilled		MSR 100% incinerated	
	I	I'	IA	IA'	IB	IB'
Collection cost ($/ton MSR)	24–	24–	24	24	24	24
Incineration cost ($/ton MSR)	8–	12–	–	–	8	12
Direct landfill cost ($/ton MSR)	3–	5–	3	5	–	––
Landfill cost ($/ton incinerator residue)	2–	4–	–	–	2	4
Total annual cost (10^6 $)	49	55	44	48	54	62

[a]The paper residuals recycled in 1969 were not included in any totals of MSR. The costs of environmental damages resulting from collection and disposal activities are not included in the above calculations.

MSR. The quantity of readily available newspapers was estimated at 100,000 tons over and above the 53,000 tons recycled in 1969. Figure 8 shows on the first line the flow of the 1.54 million tons of MSR reduced from 1.64 million tons by the additional 100,000 tons of newspapers recycled.

The 100,000 tons are divided into two different categories for estimation of collection costs that reflect the different concentrations at generation points. About 16,000 tons is designated as high-volume, readily available newspaper residuals because they are collected from high-rise apartment houses, each containing over 100 dwelling units. It is estimated that about 30 percent of the rental units in the SMSA, or 370 apartment houses in 1969, fit this category and that they generated over 300 tons of newspapers per week. Low-volume, readily available newspapers were estimated at about 84,000 tons. These residuals were generated at single-family dwellings, garden apartments, and high rises with less than 100 dwelling units. (The equation for this system by which total annual costs are calculated is shown in the appendix.)

113

The costs for SRMS-II are shown in table 13. In this table mixed solid residuals are divided into two categories, MSR collected from industrial and commercial activity and MSR collected from households. The allocation of 30 percent of municipal MSR to commercial and industrial sources and 70 percent to household sources conforms with national averages of solid residuals disposed of in municipal facilities.

The different operating costs of SRMS-II are combined in twelve variations in the table. These variations are divided into three major groups, each having four variations. The first variation in each group, II, IIA, and IIB, assumes low unit costs for incineration and landfill; the other three variations in each group assume high unit costs for incineration and/or landfill.

In each of the three major groups there are three different combinations of collection costs for household MSR and low-volume readily available newspapers. The first two variations in each group, II and II', IIA and IIA', and IIB and IIB', assume collection costs of $26.40 per ton for both household MSR and low-volume newspapers. These variations indicate the effects of varying disposal operations and varying disposal costs.

The third variation in each major disposal group, II", IIA", and IIB", assumes unit collection costs of $24 per ton for both household MSR and low-volume newspapers, reflecting the thesis that there is no increased cost in separate collection on the same truck. The fourth variation in each group, II''', IIA''', and IIB''', assumes an increase in household MSR collection cost to $26.40 while the low-volume newspaper collection cost for pickup by a separate truck is assumed to be $15 per ton.

The collection cost for commercial and industrial residuals is set in SRMS-II and all variations at $24 per ton, the same cost as used in SRMS-I. The rationale is that the commercial industrial sector of the collection operation is not affected by the separate collection of 100,000 tons of newspapers from household sources.

The collection cost of household MSR is set at $24 per ton in three of the systems and at $26.40 per ton in the other nine systems. The $24 collection cost is based on the hypothesis that separate collection of newspapers on the same truck with MSR does not delay the collection of MSR or otherwise contribute to an increased unit cost for the collection of MSR. This is the thesis presented by the city of Madison, Wisconsin, which does not show increased collection costs for MSR resulting from separate collection of newspapers.

In six systems it is hypothesized that the truck collecting newspapers is also collecting MSR; however, the collection of separated newspapers adds

114

TABLE 13

Total Annual Costs for Solid Residuals Management System II and Variations

Costs	50% Incineration 50% Landfill				100% Directly landfilled				100% Incinerated			
	II	II'	II''	II'''	IIA	IIA'	IIA''	IIA'''	IIB	IIB'	IIB''	IIB'''
Collection cost, $/ton commercial MSR	24	24	24	24	24	24	24	24	24	24	24	24
Collection cost, $/ton household MSR	26.40	26.40	24	26.40	26.40	26.40	24	26.40	26.40	26.40	24	26.40
Collection cost, $/ton $RANP_{lv}$	26.40	26.40	24	15	26.40	26.40	24	15	26.40	26.40	24	15
Collection cost, $/ton $RANP_{hv}$	10.	10.	10	10	10.	10.	10	10	10.	10.	10	10
Incineration cost, $/ton MSR	8	12	12	12	–	–	–	–	8	12	12	12
Direct landfill cost, $/ton MSR	3	5	5	5	3	5	5	5	–	–	–	–
Incinerated residue landfill cost, $/ton	2	4	4	4	–	–	–	–	2	4	4	4
Process cost, $/ton newspaper residuals	8	8	8	8	8	8	8	8	8	8	8	8
Selling price, $/ton used newspapers	22	22	22	22	22	22	22	22	22	22	22	22
Total annual cost, 10^6 $	50	55	52	54	45	48	45	47	54	62	59	61

RANP = Readily available newsprint
lv = Low volume generation
hv = High volume generation

to total pickup time, increasing the unit cost of collecting household MSR. In the remaining three systems in which the pickup cost of household MSR is posited at $26.40, the assumption is that the collection of newspapers is by a separate specialized newspaper collection truck. This leaves less weight for the MSR collection truck to pick up, but total trucks and total labor force working remain the same. Less weight collected at the same overall expense results in an increased unit cost.

The collection cost for the 84,000 tons of separate low-volume newspapers is set at $26.40 in six of the twelve variations, at $24 in three of the variations, and at $15 in three of the variations. The cost of $26.40 is hypothesized on the basis of collecting newspapers on the same truck with the household MSR but carrying them in a separate compartment. The separate compartment is assumed to be an outside carrying basket such as is used in Madison, Wisconsin.[3] The $2.40 increase per ton over the $24 figure used for MSR in the base case is assumed on the basis of additional time required to load and unload separate newspapers transported on the same truck with the household MSR.

The $24 per ton collection cost posited in II″, IIA″, and IIB″, the same as the household MSR collection costs in those variations, is based on the possible but improbable assumption that separated collections of newspapers on the regular MSR collections truck would not result in any increase at all in collection costs.

The $15 per ton collection cost for newspapers posited in variations II‴, IIA″, and IIB″ is based on collection of newspapers by a separate packer truck moving much faster than the regular collection truck and covering between two and three MSR collection routes in one shift. Assuming that the regular route trucks collect about 400 households on one route, this truck picks up 1,000 or so. This is the arrangement followed in Hempstead, Long Island, which has inaugurated separated newspaper collection.[4] As already noted, the cost of collecting household MSR in this variation is increased to $26.40 on the grounds of a smaller number of MSR spread over substantially the same MSR collection cost.

[3] Edwin Duszynski, "Solid Waste Management in Madison, Wisconsin," *Proceedings of First National Conference on Composting–Waste Recycling, Denver, Colorado, May 20–21, 1971* (Emmaus, Pa.: Rodale Press, 1971).
[4] Personal communication to author from William J. Landman, Sanitation Commissioner, Town of Hempstead, November 1, 1971.

A collection cost of $10 per ton for the high-volume, readily available newspaper residuals (amounting to 16,000 tons) generated in large high-rise apartment houses was posited in all of the variations. This cost is based on a 31-cubic yard front-end loader manned only by a driver. The newspapers are set out by apartment house personnel in portable bins of 4- or 6-cubic yard capacity. While the operating costs of the truck and the bins and labor amount to $6,000 a month, the volume of pickup is sufficiently great to yield a low cost per ton.

Incineration and landfill costs vary as in SRMS-I, as do the proportions of MSR incinerated and landfilled, but the total MSR collected is diminished by the 100,000 tons of newspapers that have been diverted to recycling and thus are not included in the costs of incineration and landfill.

The processing cost per ton of newspaper residuals, $8, covers the cost of unloading from the collection trucks, baling, storing if necessary, and loading out for shipment to a paper mill. An example of the kind of items included in this cost is the wire used for one of the large 3 X 4 X 5 bales, which costs 80 cents. The conveyor and baling equipment required to produce these 1-ton bales requires an investment upwards of $150,000.

The selling price of baled newspaper residuals used in all variations, $22, is the price quoted in the June 3, 1971, issue of *Official Board Markets* for No. 1 news, f.o.b. trucks or cars. It is at the lower end of the range of prices for this raw material.

A comparison of the total annual costs of the twelve variations of SRMS-II shows that the four variations in which all MSR are landfilled without incineration are the least expensive arrangements; next as a class are the variations in which the MSR are 50 percent landfill and 50 percent incinerated, and finally the most costly variations are the four in which all residuals are incinerated.

Within each group of incineration/landfill comparisons, the variation providing for the lower cost incineration and/or landfill is the least-cost variation. The highest cost variation in each group is that providing for a 10 percent increase in the cost of collecting MSR and newspapers with high-cost incineration and/or landfill. Among the high-cost incinerator/landfill variations, the lowest cost variation is the one providing for no increase in collection costs of either MSR or newspapers. This variation provides a somewhat lower cost than the variation in which low-volume newspaper collection goes down to $15 while household MSR collection costs are increased to $26.40.

Solid Residuals Management System III

The five variations assumed in SRMS-III provide for 60,000 more tons of used corrugated container residuals to be recycled than at present, in addition to 100,000 tons of newspaper residuals, resulting in almost a 10 percent reduction in the 1.64 million tons of MSR that go to incineration and landfill in SRMS-I.

This system is depicted in figure 9. The readily available newspaper residuals are divided into two groups, as in SRMS-II. The readily available used corrugated is likewise divided into two groups, one of low-volume generating sources yielding about 36,000 tons, and one of high-volume generating sources yielding about 24,000 tons. Two factors divide the high-volume generators from the low-volume generators. The first is the rate of generation. The high-volume generator, by definition, produces at least one 500-pound bale of corrugated cartons in one week. A 500-pound bale indicates a sales volume of about 400–425 cases of goods a week, with none of the cases being reused for packaging of retail sales, as they are in a liquor store, for example. The second factor is that the high-volume generator has a baling machine to make bales of 500 pounds or more and a truck level loading dock which allows the mechanical loading of bales.

The low-volume generator, like the high-volume generator, breaks down and bales his corrugated containers, but he does it with a manually operated baler with a capacity of 30-pound to 45-pound bales.[5] A truck level loading dock is not necessary because the bales are small enough to be loaded by hand from street level. The frequency of pickup need be no more often than every 2 weeks; thus a low-volume generator might generate as few as a dozen used corrugated cartons in a week.

The total annual costs for the variations of SRMS-III are shown in table 14. The unit costs ascribed to the various operations are the same as in SRMS-II except that two new operations have been added, the collection of readily available used corrugated residuals from low-volume generators and the collection of readily available used corrugated from high-volume generators. The collection of low-volume corrugated is posited at $24 a ton in four of the five variations and at $14 a ton in the fifth variation. The collection cost of $24 per ton for 36,000 tons of low-volume RAUC is

---

[5] An example of such a baler is Spindle-Pac described in "Study of Baling Equipment and Systems Employed in Retail Stores," by W. M. Russell, March 1970. A report prepared for the Boxboard Research and Development Association.

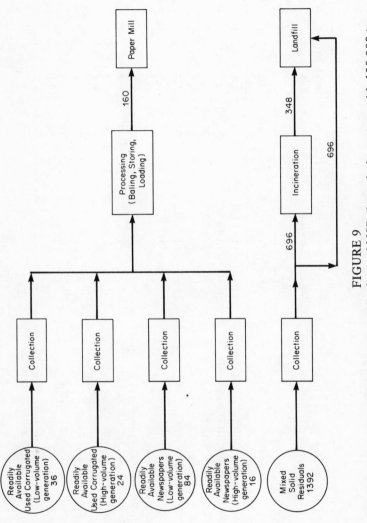

FIGURE 9

Solid residuals management system III: flow of MSR through the system with 100,000 tons of used newspapers and 60,000 tons of used corrugated containers diverted to recycling. (All figures in 1,000 tons.)

119

TABLE 14
Total Annual Costs for Solid Residuals Management System III
and Variations

Costs	Low-cost disposal, half incinerated, half landfilled		High-cost disposal, half incinerated, half landfilled		All MSR directly landfilled
	III	III$_1$	III'	III''	IIIA''
Collection cost/ton MSR (commercial)	24	24	24	24	24
Collection cost/ton MSR (household)	26.40	26.40	26.40	26.40	24
Collection cost/ton RANP$_{lv}$	26.40	26.40	26.40	26.40	24
Collection cost/ton RANP$_{hv}$	10	10.	10.	10	10
Collection cost/ton RAUC$_{lv}$[a]	24	14	24	24	24
Collection cost/ton RAUC$_{hv}$	14	24	14	14	14
Incineration cost/ton MSR	8	8	12	12.	—
Direct landfill cost/ton MSR	3	3	5	5	5
Landfill cost/ton incinerator residue	2	2	4	4.	—
Process cost/ton newspapers	8	8	8	8.	8
Process cost/ton used corrugated	13	13	13	13	13
Selling price/ton newspapers	22	22	22	22	22
Selling price/ton used corrugated	20	20	20	20	20
Total annual cost	49	49	54	51	41

[a]RAUC = Readily available used corrugated

120

based on using a straight van-type vehicle (as opposed to a packer truck or a tractor-trailer combination) with a driver and one helper. This kind of equipment has a capital cost of about one quarter of the cost of a packer truck, and an operating and labor cost 30 to 50 percent lower. It may be noted that this per ton cost coincides with the basic per ton cost for picking up MSR, and this should be identified as purely a coincidence. The per ton collection cost of this lower cost truck is driven up by the longer route it must travel to pick up a ton of low-volume RAUC, which is more widely dispersed.

The collection cost of high-volume readily available used corrugated residuals ($RAUC_{hv}$) is set in four of the five variations at \$14 per ton. This cost is based on using a tractor-trailer combination for collection, with the trailer carrying its own fork lift truck to enable mechanical loading of bales weighing 500 to 2,000 pounds. This collection operation can be carried out by a one-man crew.

The processing cost for newspaper residuals (\$8 per ton) remains the same as in SRMS-II. The cost of \$13 a ton for processing corrugated residuals is a high estimate but comes from an actual case. It covers all direct and indirect costs for unloading, breaking bales or bundles, transporting to conveyor, baling in 1,800-pound bales, storing, and loading out on trucks or railroad cars. Such items as management overhead and space are included in this figure. The selling price of \$20 is the price quoted in *Official Board Markets* for June 3, 1971, for used corrugated containers f.o.b. trucks or rail cars at dealer's plant. This is the price a paper mill would pay to a paper stock dealer.

The five columns of table 14 are divided into three groups, each with the same disposal operations and costs. The first group, consisting of III and III_1, assumes half incineration and half landfill, all at the lower unit disposal costs. The second group, III' and III'', assumes half incineration and half direct landfill but a higher unit cost than the first group. The third group, consisting only of IIIA'', assumes a 100 percent landfill disposal operation at the higher unit cost.

The only difference in the first group between III and III_1 is the reversing of the unit collection costs for the 36,000 tons of low-volume RAUC and the 24,000 tons of the high-volume RAUC to test the effect of this variation in unit costs on total annual costs. Variation III' carries the same unit costs for newspaper and MSR collection as column III but the unit disposal costs have been increased. Variation III'' assumes the same disposal costs as III' but the collection costs for household MSR and

121

low-volume newspapers are reduced. Variation IIIA″ is distinguished from III″ by having all MSR directly landfilled. In other respects it is the same as III″.

The lowest total annual cost in the SRMS-III group is variation III-A″, which provides for 100 percent landfilling without any incineration and which assumes no increase in household MSR collection costs as a result of separate newspaper pickup. It is almost 25 percent below the highest cost system shown, III′, which provides for 50 percent incineration and 50 percent landfill at higher costs and a higher collection cost for newspaper residuals and household MSR.

Comparison of the Three Systems

The accuracy of the estimated unit costs in SRMS I, II, and III is open to question. Their utility, however, does not lie in their accuracy per se, but in whether the assumptions on which they are based result in reasonably correct relative values. The most significant unit cost, because of its scale, is the cost of collecting mixed solid residuals. The commonsense case is that the separate collection of a portion of the residuals (newspapers and corrugated) will result in an overall increase in the unit cost of collection. The posited increase of 10 percent for household collection of MSR used in most of the variations is a conservatively high estimate, but it is probably correct in indicating a higher cost for collection when residuals are separated, than under the existing system.

The unit costs of disposal operations are certain to increase because of increasingly stringent environmental quality controls on incineration and landfill and also because of the scarcity of land for landfill. The estimate of a 60 percent increase in landfilling costs and a 50 percent increase in incineration costs used in most of the variations reflects the experience of many municipalities in the early 1970s.

Price data issued since the publication of the selling prices listed for used corrugated and newspaper residuals indicate that these prices may be low by 100 percent or more. All indications point to further increases in the prices of paper stock grades of newspapers and corrugated. These price increases and increases in the costs of disposal can be interpreted as encouraging for recycling. Increases in collection costs resulting from separate collection may be offset, or more than offset, by reduction of the quantity of residuals to be disposed of and by the sale of the residuals to be recycled.

The extent and trend of the system costs are indicated by a comparison of the total annual costs of the variations shown in the three different systems. The system in effect in 1969, variation I, with half the residuals incinerated, half directly landfilled, low incineration and landfill costs, and no additional recycling, is the base system with a cost of about 48 million dollars. All of the *A* variations providing for 100 percent landfill and no incineration, regardless of high or low unit operational costs, are less costly than variation I, ranging from IIA′ in which additional newspapers are recycled with increased collection costs and increased costs of landfill (about 48 million dollars) to the least expensive variation, IIIA″, providing for increased recycling of both newspaper and corrugated residuals and assuming collection costs ($24 per ton) remain the same.

The six *B* variations, which all provide for 100 percent incineration with various levels of recycling and various unit costs, are among the most costly systems, ranging from 54 million dollars for IB to over 62.5 million dollars for IB′, which provides for 100 percent incineration at the increased unit cost with no offset from the sale of newspaper or corrugated residuals. The impact of unit disposal costs on total annual costs is readily seen in the comparison of the unlettered, the *A,* and the *B* variations. The impacts are sufficiently great so that their direction is not changed by recycling. The effect of recycling on total annual costs can be gauged only by comparing the variations in SRMS I, II, and III in which operations and their unit costs are kept constant while the quantities of residuals recycled varies.

There are four comparisons which appear important in assessing the feasibility of additional recycling of paper in the Washington metropolitan area: (1) Comparison of the base line system, I, with the variation in which 100,000 tons of newspapers are recycled with increased collection costs, II, and with the additional recycling of 60,000 tons of corrugated containers, III. Disposal operations and unit costs reflect actual 1969–70 operations. (2) Comparison of base line system I′, with II″ and III″. The collection costs are held constant with the 1969 system; unit disposal costs are higher. (3) Comparison of I′ with II′ and III′. In this comparison collection costs are higher for II′ and III′. (4) Comparison of the difference in total annual cost between I′ and II″ and IB′ and IIB″ to examine the impact of recycling on systems with increased unit disposal costs.

In the first comparison, I, II, and III, unit disposal costs remain low. Collection costs for household MSR and for collection of low-volume newspapers increase by 10 percent from $24 per ton to $26.40 per ton. The total annual costs for the three variations are 49 million dollars, 49.5

123

million dollars, and 48.5 million dollars. The effect of recycling on total annual cost is not much. The increase in collection costs in II is not quite offset by the income from 100,000 tons of newspaper residuals, so there is an increase in cost of 0.5 million dollars. However, when 60,000 tons of corrugated are added to the recycling credits in III, total annual cost goes down to 48.5 million dollars, 0.5 million below the existing system. This is a net benefit resulting from recycling 160,000 tons of newspapers and corrugated in a total solid residuals load of 1.64 million tons.

In the second comparison, the cost of collecting household MSR remains the same in variations I', II'', and III'' (no increase over current collection costs). Disposal operations, 50 percent landfilled and 50 percent incinerated, remain the same, all bearing the higher unit costs. Variation II'', which provides for the recycling of 100,000 tons of newspapers, shows a total annual cost of about 52.5 million dollars, a reduction of 4.5 percent from variation I' in which no residuals are recycled. Variation III'', which provides for the recycling of 100,000 tons of newspapers and 60,000 tons of corrugated containers, shows a total annual cost of 51.5 million dollars, a reduction of 6.4 percent below the total annual cost of variation I'. These reductions in total annual costs when residuals are diverted from disposal to recycling reflect the lower disposal cost of incinerating and landfilling a smaller quantity of residuals plus the credit from the sale of the residuals diverted to recycling. Under these variations, recycling of additional quantities of newspapers and corrugated containers over and above the quantities already being recycled yields a lower total annual cost for the entire solid residuals management system. This lower total annual cost is not dependent on a higher market price for newspaper and corrugated container residuals.

The assumptions that prevail in the comparison may be challenged on the grounds that separate collection of paper residuals will cause an increase in the costs of some part of the collection operation. The effect of this proposition can be examined in the third comparison, variations I', II', and III'. The disposal costs are constant in all three variations with 50 percent landfill and 50 percent incineration at higher unit costs; however, collection costs are increased for household MSR from $24 per ton to $26.40 per ton. The collection of low-volume newspapers is also set at $26.40 per ton. Variation II' results in a total annual cost of about 55 million dollars, the same as variation I'. In other words, the credits resulting from the recycling of 100,000 tons of newspapers in II' nearly offset the increased costs of collection. When the amount of recycling is augmented in variation III′ by the addition of 60,000 tons of corrugated containers, the total annual cost goes down to about 54 million dollars or 1 million less than the variation set forth in I. Under the assumptions of this comparison, recycling comes out as a worthwhile operation even with increased collection costs.

The effect of including a recycling operation in a solid residuals management system undergoing increased costs of disposal is shown in the fourth comparison comparing the difference in total annual costs between I′ (no recycling) and II″ (recycling 100,000 tons of newspapers) with the difference in total annual costs of IB′ (no recycling) and IIB″ (recycling). Variation II″ has a lower annual cost than I′ by about 2.5 million dollars. Operations and unit costs in these variations are the same except for recycling. Variation IIB″, which includes recycling, has a lower annual cost than IB′ by about 3 million dollars. The assumptions of I′, II″ and IB′, IIB″ are parallel except that the unit cost of disposal is higher in the latter pair. In the I′ II″ comparison, residuals are 50 percent incinerated and 50 percent landfilled for an average disposal cost per ton of \$9.50 $[(12 + 5 + \frac{4}{2}) \div 2]$. In the IB′ IIB″ comparison, residuals are 100 percent incinerated at an average cost of \$14 per ton $(12 + \frac{4}{2})$ except for 100,000 tons of newspapers which are recycled. The increased cost of disposal in IB′ and IIB″ is reflected in the increased difference between recycling and no recycling (3 million dollars) as against the 2.5 million dollar difference between I′ and II″, which have the lower average disposal cost. This comparison indicates that diversion of residuals from final disposal to recycling has greater impact on systems that have higher unit disposal costs than on those with lower unit disposal costs.

Feasibility of Additional Recycling in the DC/SMSA

This study has progressed by determining the total quantity of residuals collected and disposed of in the DC/SMSA (1.64 million tons in 1969–70) and by determining that 160,000 of those tons consisted of newspapers and corrugated containers of such a quality that they could readily be recycled into new paper products. The various operations entailed in the solid residuals management system have been detailed and priced either by analogy or by estimate. A number of variations of the basic system operating in the DC/SMSA were posited and tabulated in SRMS I, II, and III.

125

These tabulations and comparisons have indicated that in the distribution of costs to the various operations of the solid residuals management system, collection operations claim by far the major share of the costs. Any action that affects the unit cost of the general collection has a substantial effect on the total annual costs of the system.

The impact of increasing unit costs of disposal operations is evident both in variations which provide for higher costs of all operations and in the variations which provide for all residuals being incinerated or all residuals being landfilled. The inevitability of increasing costs for disposal is accepted both because of increasing scarcity of suitable land for landfill and because of increasingly stringent air quality controls.

Within this framework, the effect of diverting something less than 10 percent of the total solid residuals load from disposal to recycling ranges from almost nothing to a reduction in total annual costs of about 6 percent. To the recycling ideologist this may appear an inadequate if not disappointing result. People who feel that the adequacy of natural resource supply and environmental quality in the future can be assured only through greatly increased levels of recycling seek a more dramatic demonstration of the short-range benefits of recycling than these comparisons present. But a moment's reflection should indicate that this study emerges as a rather solid demonstration on behalf of recycling. It starts with the solid residuals management system as it exists and allows, conservatively, for some changes in prices and operations that are indicated by the circumstances. The results are not contingent on technological innovation or market alteration. The recycling operation studied here could be put into effect tomorrow with positive results. In short, according to this study, the recycling of an additional 160,000 tons of newspaper and corrugated containers is economically feasible.

Appendix

Explanation

TAC	=	Total annual costs
C_c	=	Collection cost/ton MSR from commercial sources
$C_c + a$	=	Collection cost/ton for $RANP_{lv}$
$C_c + b$	=	Collection cost/ton for $RANP_{hv}$
$C_c + d$	=	Collection cost/ton MSR from households
$C_c + e$	=	Collection cost/ton for $RAUC_{lv}$

$C_c + f$ = Collection cost/ton for $RAUC_{hv}$
C_i = Incineration cost/ton
C_l = Direct landfill cost/ton mixed solid residuals
C_{li} = Landfill cost/ton incinerator residue
X = Total tons MSR collected DC/SMSA 1969 (1,640,000 tons)
$RANP_{lv}$ = Readily available newspapers from low-volume generators (84,000 tons)
$RANP_{hv}$ = Readily available newspapers from high-volume generators (16,000 tons)
Q = Total tons of mixed solid residuals (MSR) from households exclusive of $RANP_{lv}$ and $RANP_{hv}$ (1,048,000 tons)
R = Total tons mixed solid residuals from commercial/industrial sources (492,000 tons)
$RAUC_{lv}$ = Readily available used corrugated containers from low-volume generators (36,000 tons)
$RAUC_{hv}$ = Readily available used corrugated containers from high-volume generators (24,000 tons)
T = Tons of mixed solid residuals from commercial and industrial sources exclusive of $RAUC_{lv} + RAUC_{hv}$ (432,000 tons)
$$R = T + RAUC_{lv} + RAUC_{hv}$$
$$X = (Q + RANP_{lv+hv}) + (T + RAUC_{lv+hv})$$
C_{np} = Processing cost/ton newspapers
C_{uc} = Processing cost/ton used corrugated containers
SP_{np} = Selling price/ton newspapers
SP_{uc} = Selling price/ton used corrugated containers

The equations used for calculating comparative costs in SRMS I, II, and III are shown below.

SRMS-I

$$TAC = XC_c + \frac{XC_i}{2} + \frac{XC_{li}}{4} + \frac{XC_l}{2}$$

SRMS-II

$$TAC = RC_c + Q(C_c + d) + \frac{(R+Q)C_i}{2} + \frac{(R+Q)C_l}{2} + \frac{(R+Q)C_{li}}{4} + RANP_{lv}(C_c + a)$$
$$+ RANP_{hv}(C_c + b) + RANP_{lv+hv}C_{np} - RANP_{lv+hv}SP_{np}$$

127

SRMS-III

$$TAC = TC_c + Q(C_c + d) + \frac{(T+Q)C_i}{2} + \frac{(T+Q)C_l}{2} + \frac{(T+Q)C_{li}}{4} + RANP_{lv}(C_c + a)$$

$$+ RANP_{hv}(C_c + b) + RAUC_{lv}(C_c + e) + RAUC_{hv}(C_c + f) + RANP_{lv+hv}C_{np}$$

$$+ RAUC_{lv+hv}C_{uc} - RANP_{lv+hv}SP_{np} - RAUC_{lv+hv}SP_{uc}$$

7

Concluding Comments

In this exploration of the economic feasibility of diverting 160,000 tons of newspapers and corrugated containers from the 1.64 million tons of mixed solid residuals disposed of in the DC/SMSA in 1969–70, it is important to note what has not as well as what has been said about recycling. Further qualification of what has been said is necessary so the reader may appreciate the circumstances under which the conclusions of this study are valid.

A complete study of the economics of recycling entails an analysis of all of the factors involved in the use of secondary fiber as a raw material input, including those that govern the supply as well as the demand for residuals. This study has been concerned primarily with the factors and relationships affecting supply. The factors involved in demand have been indicated but not treated in detail.

The demand for used materials as raw material input relates in a variety of ways to the costs of virgin raw material, which in turn are a function of the supply and demand for this material. These factors include the costs of producing pulp, paper, and paper products, roundwood and wood products residues. The capital investment required for making a ton of paper from roundwood is much greater than the capital investment required for making a ton of paper from paper residuals. On the other hand, unit

129

operating costs of a virgin raw material paper mill are considerably lower than for a secondary materials paper mill. These factors were indicated in chapter 4 but were not developed in detail. Their existence and significance is subsumed under the market prices quoted for newspaper and corrugated container residuals and used for the cost analyses in chapter 6.

The factors that influence the supply of residuals for recycling have been considered in detail both in terms of the paper industry and in terms of the solid residuals management system. The residuals selected for study, newspapers and corrugated containers, are generated from the two largest categories of homogeneous paper products.

The necessity to discuss recycling in terms of specific paper products rather than paper generally was reviewed in chapter 4. The forty-five or forty-six grades of paper stock bearing different market prices are made up of residuals which not only have different qualities of cleanliness, but are also made from different formulations of wood pulp. Newspaper residuals and corrugated container residuals can be recycled into a variety of paper products, but these residuals carry their highest value and generally enjoy the greatest market demand when they are of a quality to be used to produce the same product from which they were generated. This case is assumed in this study. This assumption provides the basis for giving them the market values of the highest paper stock grades to which they might be assigned. It also provides a basis for assuming a market demand as great as the virgin raw material product.

Recycling paper residuals back to the products from which they were generated can be successfully carried out only when the residuals are kept separate from other residuals following generation. Some residuals can be recycled following mixed collection and separation by mechanical means. Tin cans, for example, are not seriously contaminated by mixed collection, and they can readily be separated from other residuals by a system of magnets and conveyors. Paper residuals, on the other hand, lose high-grade qualities on being mixed with other residuals and are not mechanically separable by current techniques. This means that a system for diverting newspaper and corrugated container residuals must provide for separate storage at generation site and separate collection. A basic question arises: Are the increased costs of separate collection warranted by the higher value of residuals diverted for recycling? The conclusion of this study is affirmative for the residuals in question.

There are other techniques of reclamation besides recycling to original product which are under consideration or investigation, or in some cases, in a pilot plant stage. These include composting the MSR for use as a soil conditioner; high-temperature treatment of the MSR to produce gas and fuel oil (pyrolysis); and combustion of the MSR as a supplementary fuel for the production of power plant steam. These techniques may be studied as supplements to recycling a portion of the paper residuals back to original product; that is, they would be applied to the balance of the 1.64 million tons of MSR in the DC/SMSA after the 160,000 tons of newspapers and corrugated containers had been separately collected. Or they might be considered as alternatives to paper recycling if they are economically more efficient than recycling. To be efficient, a reclamation operation would have to yield a marketable material whose processing costs minus the value of the material produced were less than the costs of disposal. For example, if residuals could be made into compost for $5 a ton and the compost marketed for $1 a ton in a system in which landfilling the residuals cost $5 a ton, it would be economically efficient to utilize composting. Thus far, the studies of composting in the United States have failed to establish a market value for compost made from MSR which would justify this technique as a supplementary means of reclaiming some value from MSR.

Pyrolysis of MSR is continuing to receive a good deal of attention and experimentation as a reclamation technique that could convert a good portion of the carbonaceous fraction of MSR into fuel. The marketability of the product would not be in question if the costs of production permitted a competitive selling price. So far, the processing costs are too high to allow this.

The combustion of MSR as a supplemental fuel in the generation of power plant steam is a reclamation technique which has been used for some time in European cities and is now in operation in three U.S. cities. With the rapid increase in the cost of fuel, there is little question that this kind of reclamation can establish itself economically as a supplementary technique. Some argue for it as the sole reclamation technique, with no provision in the management system for recycling of separated paper residuals back to original product. Others contest this position. It is estimated that "the fuel value of paper as compared to natural gas at a West Coast price of $.045 per therm is $7.20 per ton. With the rapidly spiraling upward fuel costs, it is estimated that by 1975 waste paper will have a fuel value of approximately $12 per ton. We know that waste paper as a raw material for secondary fiber plants has a present day value of $15–$60 per ton at the source of generation. . . . We are also sure that by

1975 the cost of wood and price of pulp will increase significantly, which means waste paper will still have a higher value for use as secondary fiber in paper making."[1] This quotation assumes, not unreasonably, that the price advantage of paper residuals for paper over paper residuals for fuel in the marketplace is not diluted by the relative processing costs.

Consideration of the recycling operations dealt with in this study—the recovery of newspaper and corrugated container residuals—in relation to other resource recovery operations indicates that the most efficient solid residuals management system will probably consist of a number of different kinds of recovery operations with the goal of diverting the largest quantity of high-value residuals possible from final disposal. A proper starting point would appear to be consideration of the efficient maximum recovery of each recyclable material in municipal mixed solid residuals. This would include a determination of the least net cost system (including externalities) of handling solid residuals. Such a system would consist of combinations of (1) separate handling of specific types of solid residuals from specific sources for recovery of materials to be recycled back into original products and into other products; (2) reclamation of materials from combined MSR; (3) reclamation of energy from combined MSR; and (4) various methods of final disposal of the unrecycled remainder, or all of the MSR.

Translation of the cost comparisons of SRMS I, II, and III in chapter 6 into the balance sheets of the real world encounters some pitfalls in the process. No indication is given of the variations in the distribution of benefits and costs. If a single institution were doing all of the collecting and disposing and receiving proceeds from the sale of recovered residuals as shown in the variations, the cost burden on the single institution would be expressed by the total. IN he real world of the DC/SMSA there are eight different sanitation departments collecting residuals, plus over 100 private operators. The high-volume generators of newspaper and corrugated container residuals are sources, which by current institutional arrangements, are collected by large private collectors. The low-volume generators, individual households, and small business establishments tend to be collected by the municipal governments and small private collectors. This results in unit collection costs lower for one group and higher for another group than those stated in SRMS I, II, and III. Under these

[1] Frank R. Hamilton, "The Future of the Art of Recycling Waste Paper," paper presented to TAPPI meeting, Buffalo, September 20–22, 1972.

circumstances the total net benefit may not accrue at all to one group and may be higher than stated for the other group.

No solution to this problem is suggested here except to point out that the number and variety of agents involved in solid residuals management dilute any scheme for widespread recycling of residuals in the municipal stream. Cost minimization incentives implied in SRMS I, II, and III are probably effective in this context if only one collection and disposal agent controls the entire system. As management is dispersed, incentives in some areas disappear.

The benefits and costs of diverting paper from MSR disposal for recycling would be further internalized if the residuals collection operation were integrated with the paper stock operation. In some instances there would be an opportunity for increased handling efficiency, as when loose newspapers could be delivered directly to the receiving dock of a local paper mill. In other instances the collection agency might gear its pickup operations to paper stock specifications, such as the dispatch of a truck to collect only corrugated containers from prime small generators of corrugated residuals, such as liquor stores. Emergence of a system such as this would give recycling an institutional structure and efficiency that in some part matched pulpwood operations in which the pulp cutter and marketer also has management control over the forest source.[2]

This study has sought to develop a least-cost solid residuals handling and disposal system in which an optimum (in short-run economic terms) quantity of certain paper residuals are recycled. But this system is not readily transferable to the real world, in which the solid residuals management system is highly fragmented, so that the benefits of more efficient operation and recycling might well not diminish the total cost of the system. If there is to be increased recycling and increased efficiency in the solid residuals management systems of our cities, there must be a consolidation or realignment of institutional functions which serve to internalize benefits and costs.

[2] Such a system is in fact emerging in a few locations.

133

The Johns Hopkins University Press

This book was composed in Press Roman text
and Firmin Didot Bold display type by The
Composing Room from a design by Susan
Bishop. It was printed on 55-lb. Maple
Danforth paper and bound in Joanna Arrestox
cloth by The Maple Press Company.

Library of Congress Cataloging in Publication Data

Quimby, Thomas H   E
   Recycling: the alternative to disposal.

   Includes bibliographical references.
   1. Waste paper—Washington metropolitan area—Recycling. I. Title.
TS1120.5.Q53   676′.2   74-6836
ISBN 0-8018-1655-6